R Graphics
Third Edition

Chapman & Hall/CRC
The R Series

Series Editors

John M. Chambers, Department of Statistics Stanford University Stanford, California, USA
Torsten Hothorn, Division of Biostatistics University of Zurich Switzerland
Duncan Temple Lang, Department of Statistics University of California, Davis, California, USA
Hadley Wickham, RStudio, Boston, Massachusetts, USA

Recently Published Titles

For more information about this series, please visit: https://www.crcpress.com/go/the-r-series

R Graphics
Third Edition

Paul Murrell

CRC Press
Taylor & Francis Group
Boca Raton London New York

CRC Press is an imprint of the
Taylor & Francis Group, an **informa** business

A CHAPMAN & HALL BOOK

CRC Press
Taylor & Francis Group
6000 Broken Sound Parkway NW, Suite 300
Boca Raton, FL 33487-2742

First issued in paperback 2021

© 2019 by Taylor & Francis Group, LLC
CRC Press is an imprint of Taylor & Francis Group, an Informa business

No claim to original U.S. Government works

Version Date: 20181026

ISBN-13: 978-0-367-78069-2 (pbk)
ISBN-13: 978-1-4987-8905-9 (hbk)

Library of Congress Cataloging-in-Publication Data

Names: Murrell, Paul, author.
Title: R graphics / Paul Murrell.
Description: Third edition. | Boca Raton, Florida : CRC Press, [2019] |
Series: The R series | Includes bibliographical references and index.
Identifiers: LCCN 2018040388| ISBN 9781498789059 (hardback : alk. paper) |
ISBN 9780429422768 (e-book)
Subjects: LCSH: Computer graphics. | R (Computer program language)
Classification: LCC T385 .M9 2019 | DDC 006.6/63--dc23
LC record available at https://lccn.loc.gov/2018040388

Visit the Taylor & Francis Web site at
http://www.taylorandfrancis.com

and the CRC Press Web site at
http://www.crcpress.com

Contents

II GRID GRAPHICS   121

Preface

R is a popular open-source software tool for statistical analysis and graphics. This book focuses on the graphics facilities that R provides for the production of publication-quality diagrams and plots.

What this book is about

This book describes the core graphics systems in R. The first chapter provides an overview of the R graphics facilities. There are several figures that demonstrate the variety and complexity of plots and diagrams that can be produced using R and there is a description of the overall organization of the R graphics facilities, so that the user has some idea of where to find a function for a particular purpose.

A very important feature of the R graphics setup is the existence of two distinct *graphics systems* within R: the base graphics system and the **grid** graphics system. Section 1.2.1 offers some advice on which system to use.

Part I of this book is concerned with the base graphics system, which implements many of the "traditional" graphics facilities of the S language (originally developed at Bell Laboratories and made available in a commercial implementation as S-PLUS). This system is provided by the **graphics** package. The base graphics system is older than the **grid** graphics system, and there are a greater number of graphics functions and packages written for base graphics. However, more modern **grid**-based systems, particularly **ggplot2** (see below) are now more popular. The chapters in this part of the book describe how to work with the base graphics functions, with a particular emphasis on how to modify or add output to a plot to produce exactly the right final output. Chapter 2 describes the functions that are available to produce complete plots and Chapter 3 focuses on how to customize the details of plots, combine multiple plots, and add further output to plots.

Part II describes the **grid** graphics system, which is unique to R and is more flexible than the base graphics system. The graphics facilities that are based on the **grid** graphics system are further split into three major graphics packages.

Deepayan Sarkar's **lattice** package provides a complete and coherent set of graphics functions for producing plots, based on Bill Cleveland's Trellis graphics paradigm. This is described in Chapter 4.

Hadley Wickhams' **ggplot2** package provides another complete and coherent set of graphics functions for producing plots, this time based on Leland Wilkinson's Grammar of Graphics paradigm. This is described in Chapter 5.

Finally, there is the **grid** package itself, which provides a low-level, general-purpose graphics system for producing a wide variety of images, including plots. Both **lattice** and **ggplot2** use **grid** to draw plots, but both can be used without directly encountering **grid**. The **grid** package can be used on its own, or as a low-level way to customise, modify, and combine plots produced by **lattice** or **ggplot2**. The remaining chapters in Part II describe how the **grid** system can be used to produce graphical scenes starting from a blank page. In particular, there is a discussion of how to use **grid** to develop new graphical functions that are easy for other people to use and build on.

Part III of this book is concerned with the R graphics "engine", which is provided by the **grDevices** package. This consists of functions that underly both the **graphics** and **grid** packages and provide low-level infrastructure for specifying colours and fonts (Chapter 10) and for controlling the format of R graphics output (whether we draw to the screen or save graphics to a file, such as a PDF document; see Chapter 9).

Finally, Part IV covers the integration of R graphics with other systems. Chapter 11 looks at importing graphics from other systems into R with the **grImport** and **grImport**2 packages. Chapter 12 is concerned with the problem of combining **grid** and **graphics** output using the **gridBase** and **gridGraphics** packages. Chapter 13 looks at accessing advanced graphics features that R graphics itself does not support, specifically the **gridSVG** package for producing SVG output that contains special effects like gradient fills and filter effects.

Changes in the third edition

Much of the graphics system in R that was described in the first and second editions of this book still exists and is still being heavily used, but there have been numerous changes in some of the details. One purpose of this third edition is to provide updated information on the core graphics engine, the base graphics system, and the **grid** graphics ecosystem, which includes **lattice** and **ggplot2**. In particular, Chapter 8 has been completely rewritten to accomodate changes in the recommended method for developing new **grid** grobs. The main example in that chapter has also been simplified so that it is easier to demonstrate and discuss the issues involved in developing new graphical functions and objects.

The main change from the second edition is that Part IV has been restructured. In the second edition, this part of the book attempted to cover a wide range of applications of R graphics, but the number of ways that R graphics can be used has grown to the point that Part IV would require several volumes by itself; something that is reflected in the fact that there are now many more books that cover different aspects of producing plots in R (e.g., Thomas Rahlf's *Data Visualisation with R* and Oscar Perpinan Lamigueiro's *Displaying Time Series, Spatial, and Space-time Data with R*).

The restructuring of Part IV reflects a change in focus back to static graphics in this book. For example, there is no longer a chapter on interactive graphics (which is a very active area of development, particularly connecting R with javascript libraries for use on web pages). The emphasis in this third edition is on having the ability to produce detailed and customised graphics in a wide variety of formats, on being able to share and reuse those graphics, and on being able to integrate graphics from multiple systems (for example, the chapters on importing and combining graphics in R have been retained and expanded).

As always, there is also an emphasis on producing graphics from code, with all of the benefits that accrue from that: automation, reuse, sharing, and so on.

What is different about this book?

Since the first edition of R Graphics was published, there have been many more books written about R, many of which include or even focus on producing plots with R. Notable examples include Winston Chang's *R Graphics Cookbook*, Deepayan Sarkar's *Lattice: Multivariate Data Visualization with R*, and Hadley Wickham's *ggplot2: Elegant Graphics for Data Analysis*.

One distinction with this book is that it explicitly acknowledges that the graphics facilities in R can be used to draw a wide variety of images beyond just statistical plots.

One unique feature of this book is that it provides the only comprehensive descriptions of the core graphics systems in R: base graphics and **grid** graphics.

This book focuses on the task of producing exactly the image that you want and it attempts to provide a conceptual as well as technical explanation of the steps involved. The focus is at a lower level than most other books on R graphics. There is an overview of the packages and functions that allow you to create complete plots, including **lattice** and **ggplot2**, but the value of this book is in understanding how to modify the details of those plots, add further drawing to those plots, and combine those plots with each other or make use of those plots in other systems. In addition, the knowledge in this book will allow you to draw from scratch, from the ground up, and from a blank page if that is what you need to do.

What this book is (still) not about

This book does *not* contain discussions about which sort of plot is most appropriate for a particular sort of data, nor does it contain guidelines for correct graphical presentation. In fact, instructions are provided for producing some types of plots and graphical elements that are generally disapproved of, such as pie charts and cross-hatched fill patterns.

The information in this book is meant to be used to produce a plot once the format of the plot has been decided upon and to experiment with different ways of presenting a set of data. No plot types are deliberately excluded, partly because no plot type is all bad (e.g., a pie chart can be a very effective way to represent a simple proportion) and partly because some graphical elements, such as cross-hatching, might be required by a particular publisher.

The flexibility of R graphics encourages the user *not* to be constrained to thinking in terms of just the traditional types of plots. The aim of this book is to provide lots of useful tools and to describe how to use them. There are many other sources of information on graphical guidelines and recommended plot types.

Most introductory statistics textbooks will contain basic guidelines for selecting an appropriate type of plot. Examples of books that deal specifically with the construction of effective plots and that are aimed at a general audience are *Creating More Effective Graphs* by Naomi Robbins and Edward Tufte's *Visual Display of Quantitative Information* and *Envisioning Information*. For more technical discussions of these issues, see *Graphics For Statistics and Data Analysis With R* by Kevin Keene, *Visualizing Data* and *Elements of Graphing Data* by Bill Cleveland, and *The Grammar of Graphics* by Leland Wilkinson.

For ideas on appropriate graphical displays for particular types of analysis or particular types of data, some starting points are *Data Analysis and Graphics Using R* by John Maindonald and John Braun, *An R and S-Plus Companion to Applied Regression* by John Fox, *Statistical Analysis and Data Display* by Richard Heiberger and Burt Holland, *Visualizing Categorical Data* by Michael Friendly, and *Graphical Data Analysis with R* by Antony Unwin.

This book is also *not* a complete reference to the R system. There are many freely available documents that provide both introductory and in-depth explanations of the R system. The best place to start is the "Documentation" section on the home page of the R project web site (see "On the web" on page xvi). Two examples of introductory texts are *Introductory Statistics with R* by Peter Dalgaard and *Using R for Introductory Statistics* by John Verzani.

Who should read this book

This book should be of interest to a variety of R users. For people who are new to R, this book provides an overview of the graphics facilities, which is useful for understanding what to expect from R's graphics functions and how to modify or add to the output they produce. For this purpose, Chapter 1 is the place to start. In particular, the discussion of which graphics system to use in Section 1.2.1 will be of interest. Chapters 2, 4, and 5 provide relatively brief introductions to the major packages that produce standard plots, so it should be possible to get started fairly quickly using one of those chapters.

For intermediate-level R users, this book provides all of the information necessary to perform sophisticated customizations of plots produced in R. As with many software applications, it is possible to work with R for years and remain unaware of important and useful features. This book will be useful in making users aware of the full scope of R graphics, and in providing a description of the correct model for working with R graphics. Chapters 3, 6, and 7 contain a lot of this detailed information about how R graphics works.

For advanced R users, this book contains vital information for producing coherent, reusable, and extensible graphics functions. Advanced users should pay particular attention to Chapters 6, 7, and 8.

Conventions used in this book

This book describes a large number of R functions and there are many code examples. Samples of code that could be entered interactively at the R command line are formatted as follows:

```
> 1:10
```

where the > denotes the R command-line prompt and everything else is what the user should enter. When an expression is longer than a single line it will look like the following, with the additional lines indented appropriately:

```
> plot(1:10, 1:10, col="blue", lty="dashed",
        axes=FALSE, type="l")
```

Often, the functions described in this book are used for the side effect of producing graphical output, so the result of running a function is represented by a figure. In cases where the result of a function is a value that we might be interested in, the result will be shown below the code that produced it and will be formatted as follows:

```
[1]  1  2  3  4  5  6  7  8  9 10
```

In some places, an entirely new R function is defined. Such code would normally be entered into a script file and loaded into R in one step (rather than being entered at the command line), so the code for new R functions will be presented in a figure and formatted as follows:

```
1   myfun <- function(x, y) {
2     plot(x ,y)
3   }
```

with line numbers provided for easy reference to particular parts of the code from the main text.

When referring to a function within the main text, it will be formatted in a `typewriter` font and will have parentheses after the function name, e.g., `plot()`.

When referring to the arguments to a function or the values specified for the arguments, they will also be formatted in a `typewriter` font, but they will not have any parentheses at the end, e.g., `x`, `y`, or `col="red"`.

When referring to an S3 class, statements will be of the form: "the `"classname"` class," using a typewriter font with the class name in double quotes. However, when referring to an object that is an instance of a class, statements will be of the form: "the `classname` object," using a typewriter font, but without the double quotes around the class name.

All package names are in **bold** and names of software and computer languages and formats are in Sans Serif.

On the web

There is a web site with errata and links to pages of PNG or SVG versions of all figures from the book and the R code used to produce them:

`http://www.stat.auckland.ac.nz/~paul/RG3e/`

There is also an **RGraphics** package containing functions to produce the figures in this book and all functions, classes, and methods defined in the book.

The **RGraphics** package and all other packages mentioned in this book are available from the Comprehensive R Archive Network (CRAN):

`http://cran.r-project.org/`

Version information

Software development is an ongoing process and this book can only provide a snapshot of R's graphics facilities. The descriptions and code samples in this book are accurate for R version 3.4.0, but future changes are inevitable. Much of the content of Parts I, II, and III is also accurate for earlier versions of R, but specific areas of incompatibility are not indicated in the text.

A new "minor" version of R is released every year. The most up-to-date information on the most recent versions of R and **grid** are available in the on-line help pages and at the home page for the R Project:

`http://www.R-project.org/`

Acknowledgments

One advantage of writing a new edition of this book is that it provides one of the rare opportunities to express in print my thanks to colleagues in the R Core team of developers for the work that they do to make R the powerful, reliable, and fun system that it is.

An enormous debt of gratitude is also due to the wider group of people responsible for the smooth and sane expansion of the R universe, including places like CRAN and the R-Forge web site.

Wider still is the group of enthusiastic useRs who have made the transition to developeRs and produced a staggering number of graphics extension packages for R.

Last, and most, and always, thank you, Ju.

Paul Murrell
The University of Auckland
New Zealand

This manuscript was generated on an Ubuntu 16.04 Linux system using LaTeX, numerous GNU tools, the GIMP, ghostscript, ImageMagick, **Sweave**, and many different R packages. Kudos to them all!

1

An Introduction to R Graphics

Chapter preview

This chapter provides the most basic information to get started producing plots in R. First of all, there is a three-line code example that demonstrates the fundamental steps involved in producing a plot. This is followed by a series of figures to demonstrate the range of images that R can produce. There is also a section on the organization of R graphics giving information on where to look for a particular function.

The following code provides a simple example of how to produce a plot using R (see Figure 1.1).

```
> plot(pressure)
> text(150, 600,
        "Pressure (mm Hg)\nversus\nTemperature (Celsius)")
```

The expression `plot(pressure)` produces a scatterplot of pressure versus temperature, including axes, labels, and a bounding rectangle. The call to the `text()` function adds the label at the data location (`150, 600`) within the plot.

This example is basic R graphics in a nutshell. In order to produce graphical output, the user calls a series of graphics functions, each of which produces either a complete plot or adds some output to an existing plot. R graphics follows a "painters model," which means that graphics output occurs in steps, with later output drawn on top of any previous output.

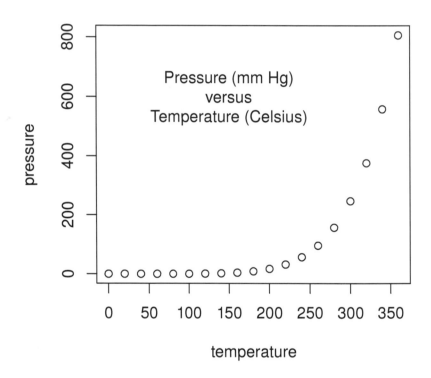

Figure 1.1
A simple scatterplot of vapor pressure of mercury as a function of temperature.
The plot is produced from two simple R expressions: one expression to draw the
basic plot, consisting of axes, data symbols, and bounding rectangle, and another
expression to add the text label within the plot.

It is also important to explicitly acknowledge that the way to produce graphics in R is by writing code. There are a number of graphical user interfaces to R that provide menus and dialog boxes for creating plots, but the only way to access the full range and power of R graphics is through code. This book takes the view that code is also the best way to produce R graphics, for several reasons: code produces a record of your actions, so that we can reproduce a plot easily, we can share the recipe for producing a plot with others, we can adapt a recipe to produce variations on a plot without having to start each time from scratch, and we can use programming tools like loops to efficiently scale up to producing large numbers of plots. Finally, code-based graphics fits nicely with version control tools, like github*, and with tools that promote reproducibility, like **knitr** and **rmarkdown**.

There are very many graphical functions provided by R and the extension packages for R so, before describing individual functions, Section 1.1 demonstrates the variety of results that can be achieved. This should provide some idea of what users can expect to be able to achieve with R graphics.

Section 1.2 gives an overview of how the graphics functions in R are organized. This should provide users with some basic ideas of where to look for a function to do a specific task. By the end of this chapter, the reader will be in a position to start understanding in more detail the core R functions that produce graphical output.

1.1   R graphics examples

This section provides an introduction to R graphics by way of a series of examples. None of the code used to produce these images is shown, but it is available from the web site for this book. The aim for now is simply to provide an overall impression of the range of graphical images that can be produced using R. The figures are described over the next few pages and the images themselves are all collected on pages 7 to 18.

1.1.1   Standard plots

R provides the usual range of standard statistical plots, including scatterplots, boxplots, histograms, barplots, pie charts, and basic 3D plots. Figure 1.2

---

*http://github.com/

shows some examples.

In R, these basic plot types can be produced by a single function call (e.g., `pie(pie.sales)` will produce a pie chart), but plots can also be considered merely as starting points for producing more complex images. For example, in the top-left scatterplot in Figure 1.2, a text label has been added within the body of the plot (in this case to show a subject identification number) and a secondary y-axis has been added on the right-hand side of the plot. Similarly, in the histogram, lines have been added to show a theoretical normal distribution for comparison with the observed data. In the barplot, labels have been added to the elements of the bars to quantify the contribution of each element to the total bar and, in the boxplot, a legend has been added to distinguish between the two data sets that have been plotted.

This ability to add several graphical elements together to create the final result is a fundamental feature of R graphics. The flexibility that this allows is demonstrated in Figure 1.3, which illustrates the estimation of the original number of vessels based on broken fragments gathered at an archaeological site: a measure of "completeness" is obtained from the fragments at the site; a theoretical relationship is used to produce an estimated range of "sampling fraction" from the observed completeness; and another theoretical relationship dictates the original number of vessels from a sampling fraction. This plot is based on a simple scatterplot, but requires the addition of many extra lines, polygons, and pieces of text, and the use of multiple overlapping coordinate systems to produce the final result.

R graphics allows fine control of very low-level aspects of a plot and these features can be used to produce some dramatic effects (at the risk of detracting from the message in the data). Figure 1.4 demonstrates one such example, where a simple barplot of tiger population levels has been embellished with an image of the head of a tiger.

For more information on the R functions that produce these standard plots, see Chapter 2. Chapter 3 describes the various ways that further output can be added to a plot.

1.1.2   Trellis plots

In addition to the base graphics plots, R provides an implementation of Trellis plots via the package `lattice` by Deepayan Sarkar. Trellis plots embody a number of design principles proposed by Bill Cleveland that are aimed at ensuring accurate and faithful communication of information via statistical plots. These principles are evident in a number of new plot types in Trellis and in the default choice of colors, symbol shapes, and line styles provided by Trellis plots. Furthermore, Trellis plots provide a feature known as *multipanel conditioning*, which creates multiple plots by splitting the data being plotted according to the levels of other variables.

Figure 1.5 shows an example of a Trellis plot. The data are yields of several different varieties of barley at six sites, over two years. The plot consists of six *panels*, one for each site. Each panel consists of a dotplot showing yield for each variety with different symbols used to distinguish different years, and a *strip* at the top showing the name of the site.

For more information on the Trellis system and how to produce Trellis plots using the **lattice** package, see Chapter 4.

1.1.3 The grammar of graphics

Leland Wilkinson's Grammar of Graphics provides another completely different paradigm for producing statistical plots and this approach to plotting has been implemented for R by Hadley Wickham's **ggplot2** package.

One advantage of this package is that it makes it possible to create a very wide variety of plots from a relatively small set of fundamental components. The **ggplot2** package also has a feature called *facetting*, which is similar to **lattice**'s multipanel plots.

Figure 1.6 shows an example of a plot that has been produced using **ggplot2**. For more information on the **ggplot2** package, see Chapter 5.

1.1.4 Specialized plots

As well as providing a wide variety of functions that produce complete plots, R provides a set of functions for producing graphical output primitives, such as lines, text, rectangles, and polygons. This makes it possible for users to write their own functions to create plots that occur in more specialized areas. There are many examples of special-purpose plots in extension packages for R. For example, Figure 1.7 shows a map of New Zealand produced using R and the extension packages **maps**, **mapdata**, and **mapproj**. Figure 1.8 shows another example: a financial chart produced by the **quantmod** package.

In some cases, researchers are inspired to produce a totally new type of plot for their data. R is not only a good platform for experimenting with novel plots, but it is also a good way to deliver new plotting techniques to other researchers. Figure 1.9 shows a novel display for decision trees, visualizing the distribution of the dependent variable in each terminal node (produced using the **party** package).

For more information on how to generate a plot starting from an empty page with base graphics functions, see Chapter 3. The **grid** package provides even more power and flexibility for producing customized graphical output (see Chapters 6 and 7), especially for the purpose of producing functions for others to use (see Chapter 8).

1.1.5   General graphical scenes

The generality and flexibility of R graphics make it possible to produce graphical images that go beyond what is normally considered to be statistical graphics, although the information presented can usually be thought of as data of some kind. A good mainstream example is the ability to embed tabular arrangements of text as graphical elements within a plot as in Figure 1.10.

R has also been used to produce figures that help to visualize important concepts or teaching points. Figure 1.11 shows two examples that provide a geometric representation of extensions to F-tests (provided by Arden Miller). It is also possible to produce flow diagrams of various sorts, as demonstrated in Figure 1.12. R graphics can even be used to produce infographics like Figure 1.13. These examples tend to require more effort to achieve the final result as they cannot be produced from a single function call.

These examples present only a tiny taste of what R graphics (and clever and enthusiastic users) can do. They highlight the usefulness of R graphics not only for producing what are considered to be standard plot types, for little effort, but also for providing tools to produce final images that are well beyond the standard plot types, including going beyond the boundaries of what is normally considered statistical graphics.

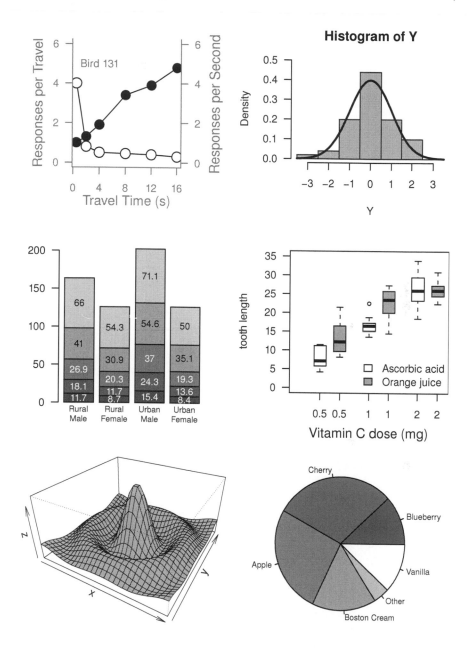

Figure 1.2

Some standard plots produced using R: (from left-to-right and top-to-bottom) a scatterplot, a histogram, a barplot, a boxplot, a 3D surface, and a pie chart. In the first four cases, the basic plot type has been augmented by adding additional labels, lines, and axes.

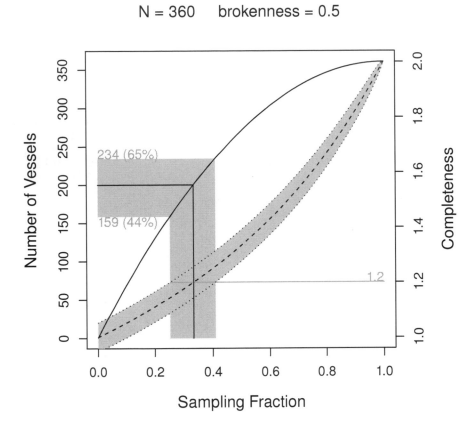

Figure 1.3
A customized scatterplot produced using R. This is created by starting with a simple
scatterplot and augmenting it by adding an additional y-axis and several additional
sets of lines, polygons, and text labels.

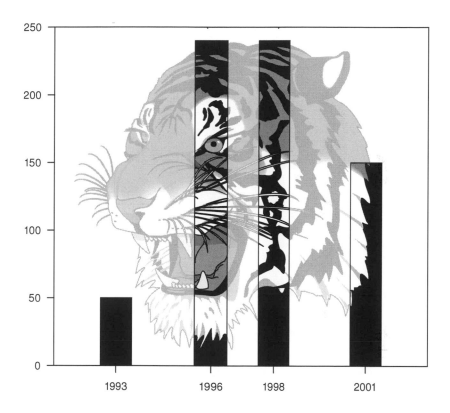

Figure 1.4
A dramatized barplot produced using R. This is created by starting with a simple barplot and augmenting it by adding a background image in light gray, with bolder sections of the image drawn in each bar.

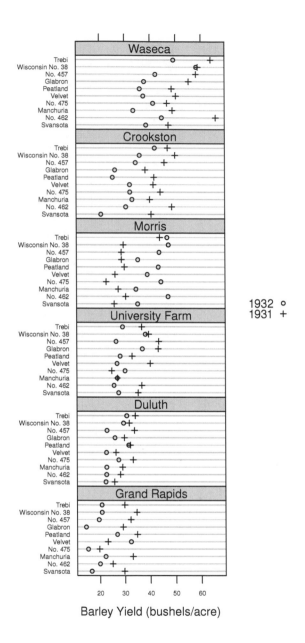

Figure 1.5

A Trellis dotplot produced using the **lattice** package. The relationship between the
yield of barley and species of barley is presented, with a separate dotplot for different
experimental sites and different plotting symbols for data gathered in different years.
This is a small modification of Figure 1.1 from Bill Cleveland's ***Visualizing Data***
(reproduced with permission from Hobart Press).

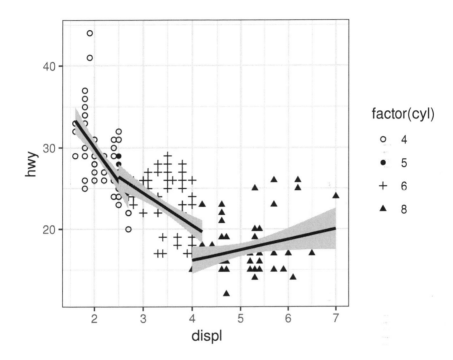

Figure 1.6
A plot produced using **ggplot2**. The relationship between miles per gallon (on the highway) and engine displacement (in liters). The data are divided into four groups based on the number of cylinders in the engine and different plotting symbols are used for each group and a separate linear model fit is shown for each group.

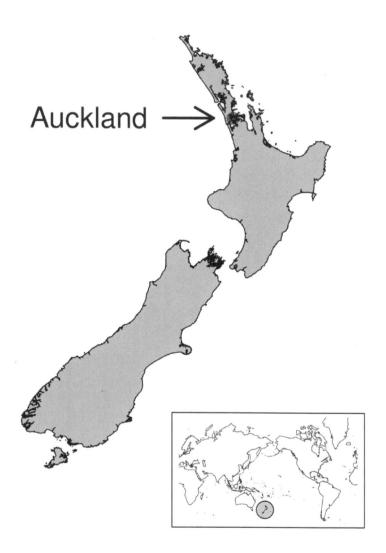

Figure 1.7
A map of New Zealand produced using the **maps** package, the **mapdata** package, and the **mapproj** package. The map (of New Zealand) is drawn as a series of polygons, and then text, an arrow, and a data point have been added to indicate the location of Auckland, the birthplace of R. A separate world map has been drawn in the bottom-right corner, with a circle to help people locate New Zealand.

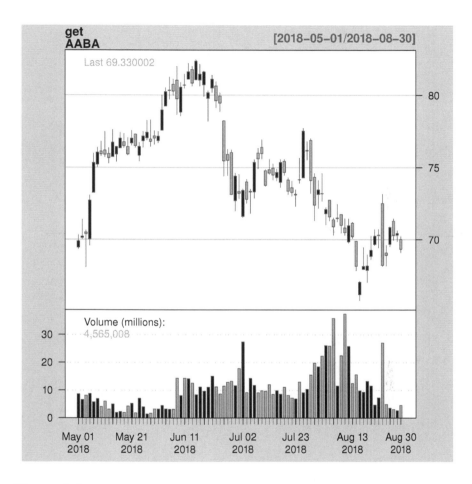

Figure 1.8
A financial chart produced with the `chartSeries()` function from the **quantmod**
package.

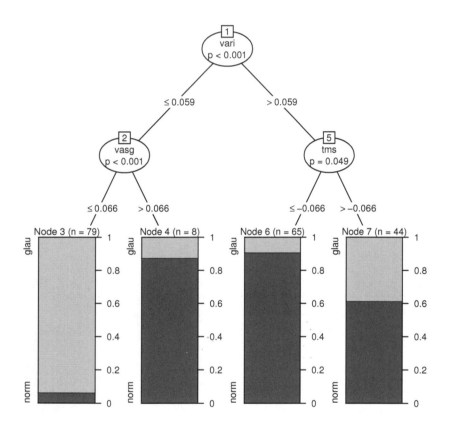

Figure 1.9
A novel decision tree plot, visualizing the distribution of the dependent variable in
each terminal node. Produced using the **party** package.

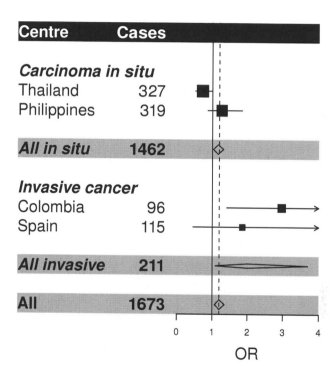

Figure 1.10
A table-like plot produced using R. This is a typical presentation of the results from a meta-analysis.

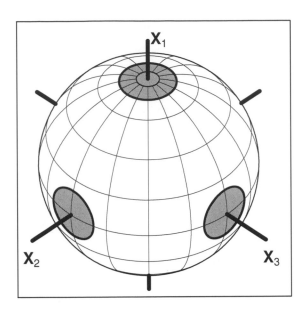

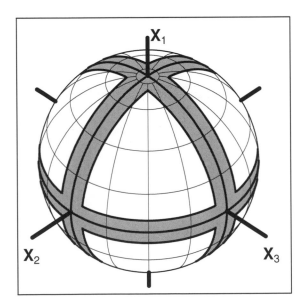

Figure 1.11
Didactic diagrams produced using R and functions provided by Arden Miller. The figures show a geometric representation of extensions to F-tests.

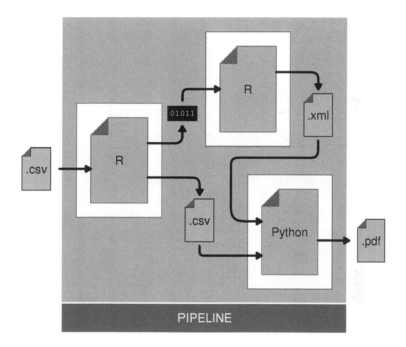

Figure 1.12

A flow diagram produced using R.

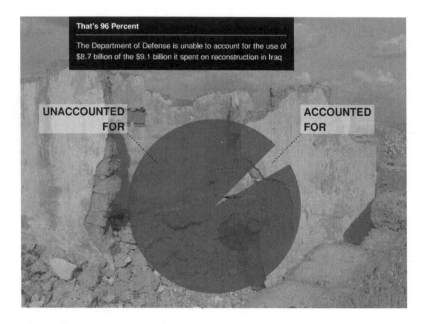

Figure 1.13
An infographic showing the proportion of aid money unaccounted for in the reconstruction of Iraq. This image is a remix of a blog post: http://www.good.is/post/infographic-where-did-the-money-to-rebuild-iraq-go/. The background image is from Adam Henning's flickr photostream: http://www.flickr.com/photos/adamhenning/66822173/.

1.2 The organization of **R** graphics

This section briefly describes how the functions and packages in the core R graphics system are organized so that the user knows where to start looking for a particular function (see Figure 1.14).

At the heart of the graphics facilities in R lies the package **grDevices**, which will be referred to as the *graphics engine*. This provides fundamental infrastucture for graphics in R, such as selecting colors and fonts and selecting a graphics output format. Although almost all graphics applications in R make use of this package, a lot can be achieved with just basic knowledge, so a detailed description of the functions in this package is delayed until Part III of this book.

Two packages build directly on top of the graphics engine: the **graphics** package and the **grid** package. These represent two largely incompatible *graphics systems* and they divide the bulk of graphics functionality in R into two separate worlds.

The **graphics** package, which will be referred to as the *base* graphics system, provides a complete set of functions for creating a wide variety of plots *plus* functions for customizing those plots in very fine detail. It is described in Part I of this book.

The **grid** package provides a separate set of basic graphics tools. It does not provide functions for drawing complete plots, so it is not often used directly to produce statistical plots. It is more common to use functions from one of the graphics packages that are built on top of **grid**, especially either the **lattice** package or the **ggplot2** package. These three packages make up the core of the **grid** graphics world in R and are described in Part II of this book.

Part IV of this book describes packages that integrate the core R graphics system either internally or with external graphics systems. The **gridBase** and **gridGraphics** packages allow output from the base graphics system and the **grid** graphics system to be combined. The **grImport** and **grImport2** packages provide tools for importing external images into R and the **gridSVG** package provides tools for adding sophisticated SVG features to **grid** graphics output.

Several other packages are also mentioned in Parts III and IV, but these are the main packages that are covered in depth.

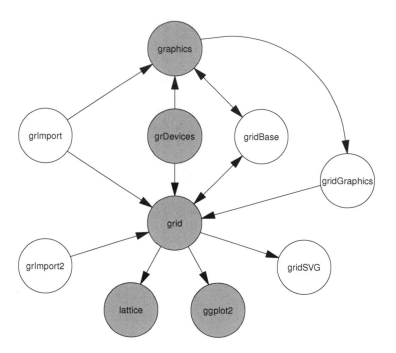

Figure 1.14

The structure of the R graphics system. The packages with gray backgrounds form the core of the graphics system. The **graphics** package is described in Part I, **grid**, **lattice**, and **ggplot2** are described in Part II, and **grDevices** is described in Part III. The packages with white backgrounds are packages that integrate the core graphics system internally or with external graphics systems and are described in Part IV.

1.2.1 Base graphics versus grid graphics

The existence of two distinct graphics systems in R, the base graphics world versus the **grid** graphics world, raises the issue of when to use each system.

For the purpose of producing complete plots from a single function call, which graphics system to use will largely depend on what type of plot is required. The choice of graphics system is largely irrelevant if no further output needs to be added to the plot.

If it is necessary to add further output to a plot, the most important thing to know is which graphics system was used to produce the original plot. In general, the same graphics system should be used to add further output (though Chapter 12 describes ways to get around this restriction).

For a wide range of standard plots, it will be possible to produce the same sort of plot in three different styles, using functions from any one of the **lattice**, **ggplot2**, or **graphics** packages. As a rough guide, the default style of the **lattice** and **ggplot2** packages may often be superior because they are both motivated by principles of human perception and designed to make it easier to extract information from a plot.

Both the **lattice** and **ggplot2** packages also provide more sophisticated support for visualizing multivariate data sets where, for example, a simple scatterplot between two continuous variables may be augmented by having separate lines or distinct plotting symbols for different subgroups within the data, or by having entire separate plots for different subgroups.

The price of the additional advanced features of both **lattice** and **ggplot2** is that there is a steeper learning curve required to master their respective conceptual frameworks. For **lattice**, there is a particular effort required to learn how to make significant customizations of the default style, while for **ggplot2**, the overall philosophy takes some getting used to, although once grasped it provides a more coherent and powerful paradigm.

In summary, given the choice, it may be quicker to get going with base graphics, but both **lattice** and **ggplot2** offer more efficient and sophisticated options in the long run.

A different problem is that of producing an image for which there is no existing function, which requires resorting to low-level graphics functions. For this situation, the **grid** system offers the benefit of a much wider range of possibilities than the low-level functions in the base graphics system, at the cost of having to learn a few additional concepts.

If the goal is to create a new graphical function for others to use, **grid** again provides better support, compared to the base graphics system, for producing more general output that can be combined with other output more easily.

One final consideration is speed. None of the core R graphics systems could be described as blindingly fast, but the **grid**-based systems are noticeably slower than base graphics and that performance penalty may be important in some applications.

Chapter summary

The R graphics system consists of a core graphics engine and two low-level graphics systems: base graphics and **grid** graphics. The base graphics system also includes high-level functions for producing complete plots. The **lattice** package and the **ggplot2** package provide high-level plotting systems on top of **grid**. Many extension packages provide further graphical facilities for both graphics systems, which means that it is possible to create a very wide range of plots and general graphical images with R.

Part I

BASE GRAPHICS

2

Simple Usage of Base Graphics

Chapter preview

This chapter introduces the main *high-level* plotting functions in the base graphics system. These are the functions used to produce complete plots such as scatterplots, histograms, and boxplots. This chapter describes the names of the standard plotting functions, the standard ways to call these functions, and some of the standard arguments that can be used to vary the appearance of the plots. Some of this information is also applicable to high-level plotting functions in extension packages.

The aim of this chapter is to provide an idea of the range of plots that are available in the base graphics system, to point the user toward the most important ones, and to introduce the standard approach to using them.

Although the focus of this book is on controlling the fine details of plots, we must first have a plot to fine tune. This chapter describes how to generate a range of complete plots within the base graphics system.

The graphics functions that make up the base graphics system are provided in an extension package called **graphics**, which is automatically loaded in a standard installation of R. In a non-standard installation, it may be necessary to make the following call in order to access base graphics functions (if the **graphics** package is already loaded, this will not do any harm).

```
> library(graphics)
```

This chapter mentions many of the high-level graphics functions in the **graphics** package, but does not describe all possible uses of these functions. For

detailed information on the behavior of individual functions the user will need
to consult the individual help pages using the `help()` function. For example,
the following code shows the help page for the `barplot()` function.

```
> help(barplot)
```

Another useful way of learning about a graphics function is to use the
`example()` function. This runs the code in the "Examples" section of the help
page for a function. The following code runs the examples for `barplot()`.

```
> example(barplot)
```

2.1   The base graphics model

As described at the start of Chapter 1, a plot is created in base graphics by
first calling a high-level function that creates a complete plot, then calling
low-level functions to add more output if necessary.

If there is only one plot per page, then a high-level function starts a new plot
on a new page. There may be multiple plots on a page, in which case a high-
level function starts the next plot on the same page, only starting a new page
when the number of plots per page is exceeded (see Section 3.3). All low-level
functions add output to the current plot. It is not generally possible to go
back to a previous plot in the base graphics system (see Section 3.3.3 for an
exception).

2.2   The `plot()` function

The most important high-level function in base graphics is the `plot()` function. In many situations, this provides the simplest way to produce a complete
plot in R.

The first argument to `plot()` provides the data to plot and there is a reasonable amount of flexibility in the way that the data can be specified. For
example, each of the following calls to `plot()` can be used to produce essentially the same scatterplot (shown in Figure 2.1), with small variations in the

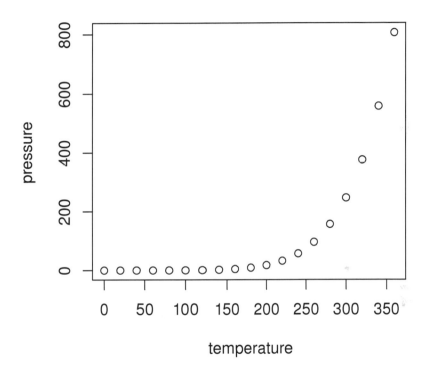

Figure 2.1
A scatterplot produced by the `plot()` function. This plot can be produced by providing a single data frame, two numeric vectors, or a formula as the first argument to the `plot()` function.

axis labels. In the first case, all of the data to plot are specified in a single data frame. In the second case, separate x and y variables are specified as two separate arguments. In the third case, the data to plot are specified as a formula of the form y ~ x, plus a data frame that contains the variables mentioned in the formula.

```
> plot(pressure)
> plot(pressure$temperature, pressure$pressure)
> plot(pressure ~ temperature, data=pressure)
```

Base graphics does not make a major distinction between, for example, scatterplots that only plot data symbols at each (x, y) location and scatterplots that draw straight lines connecting the (x, y) locations (line plots). These are just variations on the basic scatterplot, controlled by a type argument. This is demonstrated by the following code, which produces four different plots by varying the value of the type argument (see Figure 2.2).

```
> plot(pressure, type="p")
> plot(pressure, type="l")
> plot(pressure, type="b")
> plot(pressure, type="h")
```

Base graphics also does not make a distinction between a plot of a single set of data and a plot containing multiple series of data. Additional data series can be added to a plot using low-level functions such as points() and lines() (see Section 3.4.1; also see the function matplot() in Section 2.5).

The plot() function is *generic*. One consequence of this has just been described; the plot() function can cope with the same data being specified in several different *formats* (and it will produce the same result). However, the fact that plot() is generic also means that if plot() is given different *types* of data, it will produce different types of plots. For example, the plot() function will produce boxplots, rather than a scatterplot, if the x variable is a factor, rather than a numeric vector. Another example is shown in the code below. Here an "lm" object is created from a call to the lm() function. When this object is passed to the plot() function, the special plot method for "lm" objects produces several regression diagnostic plots (see Figure 2.3).*

```
> lmfit <- lm(sr ~ pop15 + pop75 + dpi + ddpi,
              data = LifeCycleSavings)
> plot(lmfit)
```

---

*The data used in this example are measures relating to the savings ratio (aggregate personal saving divided by disposable income) averaged over the period 1960-1970 for 50 countries, available as the data set LifeCycleSavings in the datasets package.

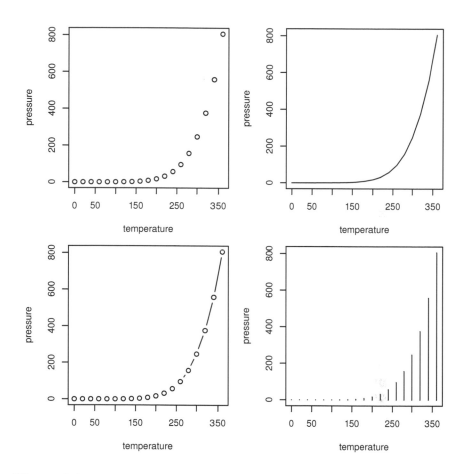

Figure 2.2

Four variations on a scatterplot. In each case, the plot is produced by a call to the `plot()` function with the same data; all that changes is the value of the `type` argument. At top-left, `type="p"` to give **p**oints (data symbols), at top-right, `type="l"` to give **l**ines, at bottom-left, `type="b"` to give **b**oth, and at bottom-right, `type="h"` to give **h**istogram-like vertical lines.

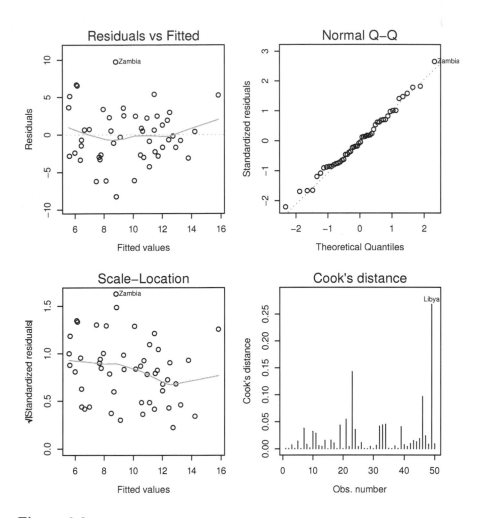

Figure 2.3
Plotting an "lm" object. There is a special plot() method for "lm" objects that
produces a number of diagnostic plots from the results of a linear model analysis.

In order to learn more about the "lm" method for the plot() function, type
help(plot.lm).

In many cases, graphics extension packages provide new types of plots by
defining a new method for the plot() function. For example, the **cluster**
package provides a plot() method for plotting the result of an agglomerative
hierarchical clustering procedure (an **agnes** object). This method produces a
special bannerplot and a dendrogram from the data (see the following code
and Figure 2.4).* The first block of expressions is just setting up the data
and creating an **agnes** object; the last expression plots the **agnes** object.

```
> subset <- sample(1:150, 20)
> cS <- as.character(Sp <- iris$Species[subset])
> cS[Sp == "setosa"] <- "S"
> cS[Sp == "versicolor"] <- "V"
> cS[Sp == "virginica"] <- "g"
> ai <- agnes(iris[subset, 1:4])

> plot(ai, labels = cS)
```

Simply calling plot(x), where x is an R object containing the data to visualize,
is often the simplest way to get an initial view of the data.

The following sections briefly describe the main types of plots that can be
produced using either plot() or one of the other high-level functions in the
graphics package. Toward the end of the chapter is a discussion of important arguments to these functions that allow some control over the detailed
appearance of the plots (see Section 2.6).

2.3   Plots of a single variable

Table 2.1 and Figure 2.5 show the base graphics functions that produce a plot
based on a single variable.

The plot() function will accept, as a single unnamed argument, a numeric
vector, or a factor, or a one-dimensional table (a table of counts from a single

---

*The data used in this example are the famous iris data set giving measurements of
physical dimensions of three species of iris, available as the iris data set in the **datasets**
package.

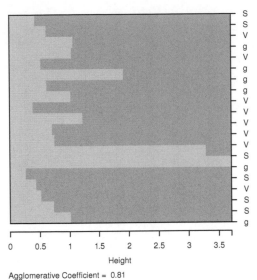

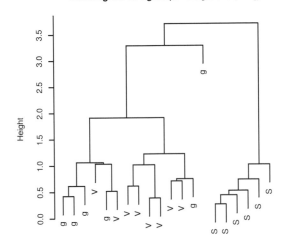

Figure 2.4

Plotting an **agnes** object. There is a special `plot()` method for **agnes** objects that produces plots relevant to the results of an agglomerative hierarchical clustering analysis.

Table 2.1

High-level base graphics plotting functions for producing plots of a single variable.

Function	Data	Description
plot()	Numeric	Scatterplot
plot()	Factor	Barplot
plot()	1-D table	Barplot
barplot()	Numeric (bar heights)	Barplot
pie()	Numeric	Pie chart
dotchart()	Numeric	Dotplot
boxplot()	Numeric	Boxplot
hist()	Numeric	Histogram
stripchart()	Numeric	1-D scatterplot
stem()	Numeric	Stem-and-leaf plot

factor). A numeric vector will produce a scatterplot of the numeric values as a function of their indices, while both a factor and a table produce a barplot of the counts for each level of the factor. The plot() function will also accept a formula of the form ˜ x and if the variable x is numeric, the result is a one-dimensional scatterplot (stripchart). If x is a factor, the result is a barplot.

A barplot can also be produced explicitly with the barplot() function. The difference is that this function requires a numeric vector, rather than a factor, as input — the numeric values are treated as the heights of the bars to be plotted.

One issue with producing a barplot is providing a meaningful label below each bar. The plot() function uses the levels of the factor being plotted for bar labels and barplot() will use the **names** attribute of the numeric vector if it is available.

As alternatives to a barplot, the pie() function plots the values in a numeric vector as a pie chart, and dotchart() produces a dotplot.

Several functions provide a variety of ways to view the distribution of values in a single numeric vector. The boxplot() function produces a boxplot (or box-and-whisker plot), the hist() function produces a histogram, stripchart() produces a one-dimensional scatterplot (stripchart), and stem() produces a stem-and-leaf plot (but as text, on the console, rather than graphical output).

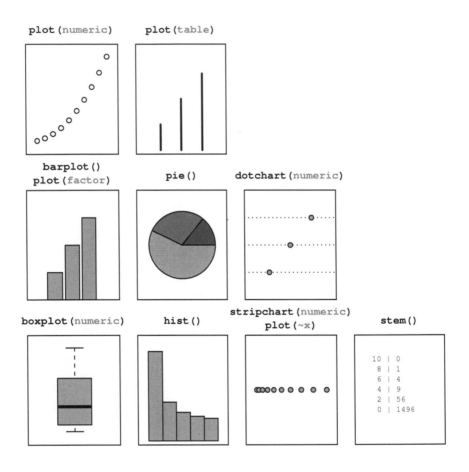

Figure 2.5
High-level base graphics plotting functions for producing plots of a single variable.
Where the function can be used to produce more than one type of plot, the relevant
data type is shown (in gray). For example, **plot**(numeric) means that this is what
the **plot**() produces when it is given a single numeric argument.

2.4   Plots of two variables

Table 2.2 and Figure 2.6 show the base graphics functions that produce plots of two variables.

The `plot()` function will accept two variables in a variety of formats: a pair of numeric vectors; one numeric vector and one factor; two factors; a list of two vectors or factors (named `x` and `y`); a two-dimensional table; a matrix or data frame with two columns (the first column is treated as `x`); or a formula of the form `y ~ x`.

If both variables are numeric, the result is a scatterplot. If `x` is a factor and `y` is numeric, the result is a boxplot for each level of `x`. If `x` is numeric and `y` is a factor, the result is a (grouped) stripchart, and if both variables are factors, the result is a spineplot. If `plot()` is given a table of counts, the result is a mosaic plot.

Two functions provide alternatives to the scatterplot, both motivated by the problem of overplotting, which occurs when values repeat or when there are very many points to plot. The `sunflowerplot()` function draws a special symbol at each location to indicate how many points are overplotted and the `smoothScatter()` function draws a representation of the density of points in the scatterplot (rather than drawing individual points). Another way to produce multiple stripcharts is to provide `stripchart()` with a list of numeric vectors.

When `x` is a factor and `y` is numeric, another way to produce multiple boxplots is with the `boxplot()` function, with the data provided either as a list of numeric vectors or as a formula of the form `y ~ x`, where `x` is a factor.

If the data consist of a numeric matrix, where each column or row represents a different group, the `barplot()` function will produce a stacked or side-by-side barplot from the numeric values and `dotchart()` will produce a dotplot.

When `x` is numeric and `y` is a factor, the `spineplot()` function will produce a spinogram, and `cdplot()` will produce a conditional density plot. Both functions will also accept the data as a formula of the form `y ~ x`.

For plotting two factors, there are also several options. Given the raw factors, the `spineplot()` function will produce a spineplot, just like `plot()` produces from two factors. An alternative is to work with a table of counts of the two factors. Given a table, the `mosaicplot()` function produces a mosaic plot, just like `plot()` does. The `mosaicplot()` function will also accept a formula of the form `y ~ x` where both `y` and `x` are factors.

Table 2.2

High-level base graphics plotting functions for producing plots of two variables.

Function	Data	Description
plot()	Numeric, numeric	Scatterplot
plot()	Numeric, factor	Stripcharts
plot()	Factor, numeric	Boxplots
plot()	Factor, factor	Spineplot
plot()	2-D table	Mosaic plot
sunflowerplot()	Numeric, numeric	Sunflower scatterplot
smoothScatter()	Numeric, numeric	Smooth scatterplot
boxplot()	List of numeric	Boxplots
barplot()	Matrix	Stacked/side-by-side barplot
dotchart()	Matrix	Dotplot
stripchart()	List of numeric	Stripcharts
spineplot()	Numeric, factor	Spinogram
cdplot()	Numeric, factor	Conditional density plot
fourfoldplot()	2x2 table	Fourfold display
assocplot()	2-D table	Association plot
mosaicplot()	2-D table	Mosaic plot

In the special case where both factors have only two levels, `assocplot()` produces a Cohen-Friendly association plot and `fourfoldplot()` produces a fourfold display.

In addition to the numeric vector and factor data types, another important basic data type is *dates* (or *date-times*). If `plot()` is given either x or y as a `"Date"` or `"POSIXt"` object, then the corresponding axis will be labeled with date descriptions (e.g., using month names).

2.5   Plots of many variables

Table 2.3 and Figure 2.7 show the base graphics functions that produce plots of many variables.

Given a numeric data frame, the `plot()` function will produce a scatterplot matrix, plotting all pairs of variables against each other. The `pairs()` function does likewise, but it will accept the data in matrix form as well.

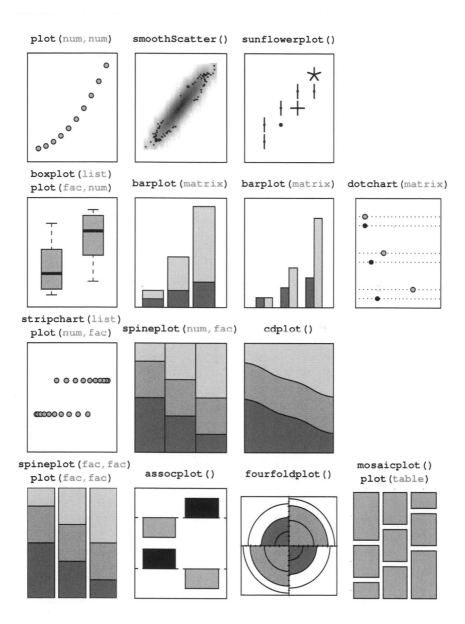

Figure 2.6
High-level base graphics plotting functions for producing plots of two variables.
Where the function can be used to produce more than one type of plot, the relevant
data type is shown (in gray). For example `plot(num,fac)` represents calling the
`plot()` function with a numeric vector as the first argument and a factor as the
second argument.

Table 2.3
High-level base graphics plotting functions for producing plots of many variables.

Function	Data	Description
plot()	Data frame	Scatterplot matrix
pairs()	Matrix	Scatterplot matrix
matplot()	Matrix	Scatterplot
stars()	Matrix	Star plots
image()	Numeric,numeric,numeric	Image plot
contour()	Numeric,numeric,numeric	Contour plot
filled.contour()	Numeric,numeric,numeric	Filled contour
persp()	Numeric,numeric,numeric	3-D surface
symbols()	Numeric,numeric,numeric	Symbol scatterplot
coplot()	Formula	Conditioning plot
mosaicplot()	*N*-D table	Mosaic plot

An alternative, when the data are in matrix form, is the matplot() function, which will plot a single scatterplot with a separate series of data symbols or lines for each column of data. The data can be separate x and y matrices, or a single matrix, in which case the values are treated as y-values and plotted against 1:nrow.

Another alternative is the stars() function, which draws a star for each row of data, with the values in the columns columns dictating the lengths of the arms of each star. This type of plot is an example of the *small multiples* technique, where many small plots are produced on a single page (see Section 3.3 for details on how to place multiple plots of any sort on a single page).

Several functions cater for the special case of three numeric variables. When x and y are measured on a regular grid, and there is a single response variable, z, the image() function plots z as a grid of colored regions, the contour() function draws contour lines (lines of constant z), filled.contour() produces colored regions *between* contour lines, and persp() produces a three-dimensional surface to represent z.

The symbols() function produces a scatterplot of x and y with a small symbol used to represent z, for example, a circle with radius proportional to z. A range of symbols is provided, some of which allow multiple variables to be represented within the symbol, for example, a rectangle symbol can encode separate variables as the width and height of the rectangle.

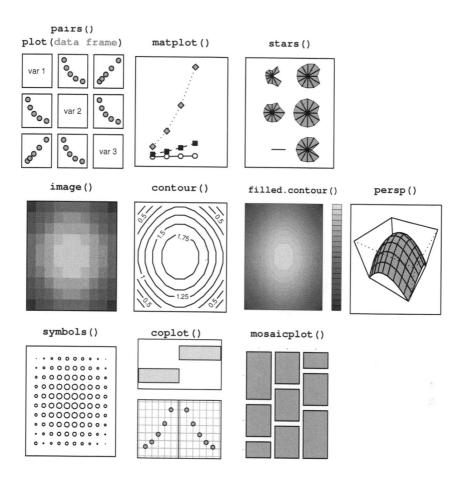

Figure 2.7
High-level base graphics plotting functions for producing plots of many variables.
Where the function can be used to produce more than one type of plot, the relevant
data type is shown (in gray).

When the data consist of two numeric variables and one or two grouping factors, the `coplot()` function can be used to produce a conditioning plot, which draws a separate plot for each level of the grouping factors. The data must be given to this function as a formula of the form y ~ x | g or y ~ x | g*h, where g and h are factors. This idea is implemented on a much grander scale in the **lattice** package (see Chapter 4) and in the **ggplot2** package (see Chapter 5).

For data consisting of multiple factors, the `mosaicplot()` function will produce a multidimensional mosaic plot, given a multidimensional table of counts.

2.6   Arguments to graphics functions

It is often the case, especially when producing graphics for publication, that the output produced by a single call to a high-level graphics function is not exactly right in all its details. There are many ways in which the output of graphics functions may be modified and Chapter 3 addresses this topic in full detail. This section will only consider the possibility of specifying arguments to high-level graphics functions in order to modify their output.

Many of these arguments are specific to a particular function. For example, the `boxplot()` function has `width` and `boxwex` arguments (among others) for controlling the width of the boxes in the plot, and the `barplot()` function has a `horiz` argument for controlling whether bars are drawn horizontally rather than vertically. The following code shows examples of the use of the `boxwex` argument for `boxplot()` and the `horiz` argument for `barplot()` (see Figure 2.8).

In the first example, there are two calls to `boxplot()`, which are identical except that the second specifies that the individual boxplots should be half as wide as they would be by default (`boxwex=0.5`).*

```
> boxplot(decrease ~ treatment, data = OrchardSprays,
          log = "y", col="light gray")
> boxplot(decrease ~ treatment, data = OrchardSprays,
          log = "y", col="light gray",
          boxwex=0.5)
```

---

*The data used in this example are amount of orchard spray consumed by honey bees for different potencies, available as the `OrchardSprays` data frame from the **datasets** package.

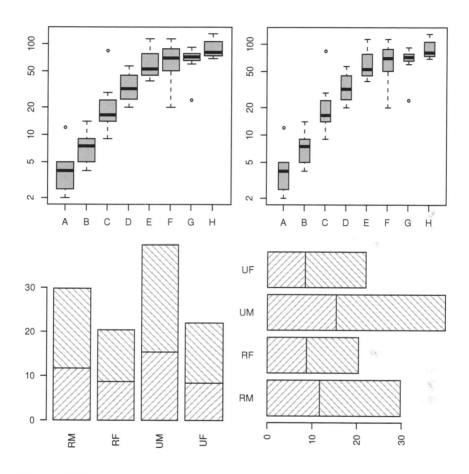

Figure 2.8

Modifying default `barplot()` and `boxplot()` output. The top two plots are produced by calls to the `boxplot()` function with the same data, but with different values of the `boxwex` argument. The bottom two plots are both produced by calls to the `barplot()` function with the same data, but with different values of the `horiz` argument.

In the second example, there are two calls to `barplot()`, which are identical except that the second specifies that the bars should be drawn horizontally rather than vertically (`horiz=TRUE`).*

```
> barplot(VADeaths[1:2,], angle = c(45, 135),
          density = 20, col = "gray",
          names=c("RM", "RF", "UM", "UF"))
> barplot(VADeaths[1:2,], angle = c(45, 135),
          density = 20, col = "gray",
          names=c("RM", "RF", "UM", "UF"),
          horiz=TRUE)
```

In general, the user should consult the documentation for a specific function to determine which arguments are available and what effect they have.

2.6.1   Standard arguments to graphics functions

Despite the existence of many arguments that are specific only to a single graphics function, there are several arguments that are "standard" in the sense that many high-level base graphics functions will accept them.

Most high-level functions will accept *graphical parameters* that control such things as color (`col`), line type (`lty`), and text font (`font` and `family`). Section 3.2 provides a full list of these arguments and describes their effects. Chapter 10 describes the complete set of values that these arguments can take.

Unfortunately, because the interpretation of these standard arguments may vary in some cases, some care is necessary. For example, if the `col` argument is specified for a standard scatterplot, this only affects the color of the data symbols in the plot (it does not affect the color of the axes or the axis labels), but for the `barplot()` function, `col` specifies the color for the fill or pattern used within the bars.

In addition to the standard graphical parameters, there are standard arguments to control the appearance of axes and labels on plots. It is usually possible to modify the range of the axis scales on a plot by specifying `xlim` or `ylim` arguments in the call to the high-level function, and often there is a set of arguments for specifying the labels on a plot: `main` for a title, `sub` for a subtitle, `xlab` for an x-axis label and `ylab` for a y-axis label.

---

*The data used in this example are death rates in Virginia in 1940 broken into different age groups and by gender and rural/urban location, available as the **VADeaths** matrix from the **datasets** package.

Although there is no guarantee that these standard arguments will be accepted by high-level functions in graphics extension packages, in many cases they will be accepted, and they will have the expected effect.

The following code shows examples of setting some of these standard arguments for the `plot()` function (see Figure 2.9). All of the calls to `plot()` draw a scatterplot of the same data with lines connecting the data values: the first call uses a wider line (`lwd=3`), the second call draws the line a gray color (`col="gray"`), the third call draws a dashed line (`lty="dashed"`), and the fourth call uses a much wider range of values on the y-scale (`ylim=c(-4, 4)`).

```
> y <- rnorm(20)
> plot(y, type="l", lwd=3)
> plot(y, type="l", col="gray")
> plot(y, type="l", lty="dashed")
> plot(y, type="l", ylim=c(-4, 4))
```

In cases where the default output from a high-level function cannot be modified to produce the desired result by just specifying arguments to the high-level function, possible options are to add further output to the plot using low-level graphics functions (see Section 3.4), or to generate the entire plot from scratch (see Section 3.5).

Some high-level functions provide an argument to inhibit some of the default output in order to assist in the customization of a plot. For example, the default `plot()` function has an **axes** argument to allow the user to inhibit the drawing of axes and an **ann** argument to inhibit the drawing of axis labels; the user can then produce customized output to represent the axes and labels (see Section 3.4.4).

2.7 Specialized plots

The base graphics system, and the extension packages that are built on it, contain a number of functions to produce plots that are suited to a particular type of data or analysis technique, or that are specific to a particular area of research.

Several of these are just variations on a basic scatterplot, with data symbols and/or lines plotted on cartesian coordinates. For example, the `qqplot()` and `qqnorm()` functions produce quantile-quantile plots (plotting observed values

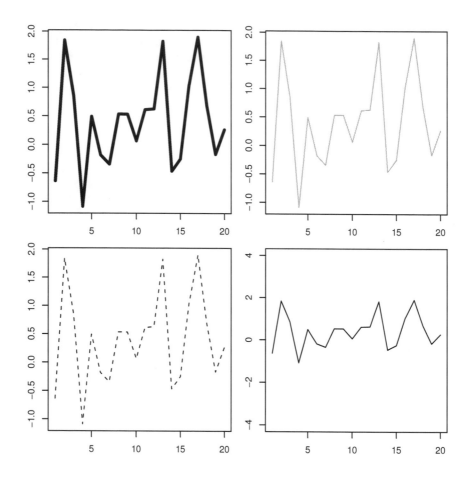

Figure 2.9

Standard arguments for high-level functions. All four plots are produced by calls to the plot() function with the same data, but with different standard plot function arguments specified: the top-left plot makes use of the lwd argument to control line thickness; the top-right plot uses the col argument to control line color; the bottom-left plot makes use of the lty argument to control line type; and the bottom-right plot uses the ylim argument to control the scale on the y-axis.

against values generated from theoretical distributions), the `plot()` method
for `"ecdf"` objects (empirical cumulative distribution functions) draws a step
plot, and the `plot()` methods for `"ts"` (time series) objects or density esti-
mates (from the `density()` function) automatically draw lines between values
to show the appropriate trends.

One interesting case is the display of a parametric curve where, rather than
specifying explicit data points, a *relationship* between x and y is provided.
This can be achieved in two ways: via the `plot()` method for function objects
and via the `curve()` function. The following code shows both approaches to
draw a sine wave (see Figure 2.10). In the first case, we must provide a function
as the first argument to the `plot()` function, but in the second case, we can
just provide an expression as the first argument to the `curve()` function.

```
> plot(function(x) {
          sin(x)/x
      },
      from=-10*pi, to=10*pi,
      xlab="", ylab="", n=500)


> curve(sin(x)/x, -10*pi, 10*pi)
```

There are also some functions that produce quite different sorts of plots. The
`plot()` method for `dendrogram` objects is provided for drawing hierarchical
or tree-like structures, such as the results from clustering or a recursive partitioning regression tree. The bottom two plots in Figure 2.10 show examples
of output from the `plot()` method for `dendrogram` objects.*

---

*The data used in these examples are measures of crime rates in various US states in
1973, available as the data set `USArrests` in the `datasets` package.

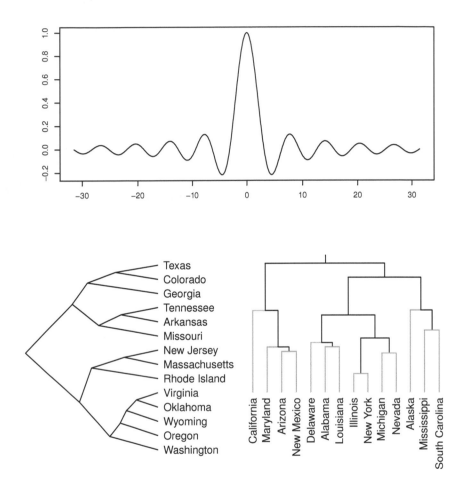

Figure 2.10
Some specialized plots. At the top is a plot of an R function and along the bottom
are two variations on a dendrogram.

Chapter summary

The base graphics system has functions to produce the standard statistical plots such as histograms, scatterplots, barplots, and pie charts. There are also functions for producing higher-dimensional plots such as 3D surfaces and contour plots and more specialized or modern plots such as dotplots, dendrograms, and mosaic plots. In most cases, the functions provide a number of arguments to allow the user to control the details of the plot, such as the widths of the boxes in a boxplot. There is also a standard set of arguments for controlling the appearance of a plot, such as colors, fonts, and line types and axis ranges and labeling, although these are not all available for all types of plots.

3

Customizing Base Graphics

Chapter preview

It is very often the case that a high-level plotting function does not produce exactly the final result that is desired. This chapter describes *low-level* base graphics functions that are useful for controlling the fine details of a plot and for adding further output to a plot (e.g., adding descriptive labels).

In order to utilize these low-level functions effectively, this chapter also includes a description of the regions and coordinate systems that are used to locate the output from low-level functions. For example, there is a description of which function to use to draw text in the margins of a plot as opposed to drawing text in the data region (where the data symbols are plotted). There is also a discussion of ways to arrange several plots together on a single page.

Sometimes it is not possible to achieve a final result by modifying an existing high-level plot. In such cases, the user might need to create a plot using only low-level functions. This case is also addressed in this chapter together with some discussion of how to write a new graphics function for other people to use.

It is often the case that the default or standard output from a high-level function is not exactly what the user requires, particularly when producing graphics for publication. Various aspects of the output often need to be modified or completely replaced. This chapter describes the various ways in which the output from a base graphics high-level function can be customized and extended.

The real power of the base graphics system lies in the ability to control many aspects of the appearance of a plot, to add extra output to a plot, and even to build a plot from scratch in order to produce precisely the right final output.

Section 3.1 introduces important concepts of *drawing regions, coordinate systems*, and *graphics state* that are required for properly working with base graphics at a lower level. Section 3.2 describes how to control aspects of output such as colors, fonts, line styles, and plotting symbols, and Section 3.3 addresses the problem of placing several plots on the same page. Section 3.4 describes how to customize a plot by adding extra output and Section 3.5 looks at ways to develop entirely new types of plots.

3.1 The base graphics model in more detail

In order to explain some of the facilities for customizing plots, it is necessary to describe more about the model underlying base graphics plots.

3.1.1 Plotting regions

In the base graphics system, every page is split up into three main regions: the *outer margins*, the current *figure region*, and the current *plot region*. Figure 3.1 shows these regions when there is only one figure on the page and Figure 3.2 shows the regions when there are multiple figures on the page.

The region obtained by removing the outer margins from the device is called the *inner region*. When there is only one figure, this usually corresponds to the figure region, but when there are multiple figures the inner region corresponds to the union of all figure regions.

The area outside the plot region, but inside the figure region is referred to as the *figure margins*. A typical high-level function draws data symbols and lines within the plot region and axes and labels in the figure margins or outer margins (see Section 3.4 for information on the functions used to draw output in the different regions). The margins are numbered 1 to 4 in the order bottom, left, top, then right. For example, "margin 3" means the top margin.

The size and location of the different regions are controlled either via the `par()` function, or using special functions for arranging plots (see Section 3.3). Specifying an arrangement of the regions does not usually affect the current plot as the settings only come into effect when the next plot is started.

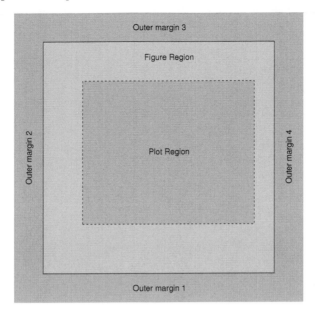

Figure 3.1
The plot regions in base graphics — the outer margins, figure region, and plot region — when there is a single plot on the page.

Coordinate systems

Each plotting region has one or more coordinate systems associated with it. Drawing in a region occurs relative to the relevant coordinate system. The coordinate system in the plot region, referred to as *user coordinates*, is probably the easiest to understand as it simply corresponds to the range of values on the axes of the plot (see Figure 3.3). The drawing of data symbols, lines, and text in the plot region occurs relative to this user coordinate system.

The scales on the axes of a plot are often set up automatically by R, but Sections 2.6 and 3.4.4 describe ways to set the scales manually.

The figure margins contain the next most commonly used coordinate systems. The coordinate systems in these margins are a combination of x- or y-ranges (like user coordinates) and lines of text away from the boundary of the plot region. Figure 3.4 shows two of the four figure margin coordinate systems. Axes are drawn in the figure margins using these coordinate systems.

There is a further set of "normalized" coordinate systems available for the figure margins in which the x- and y-ranges are replaced with a range from 0 to 1. In other words, it is possible to specify locations along the axes as a proportion of the total axis length. Axis labels and plot titles are drawn relative

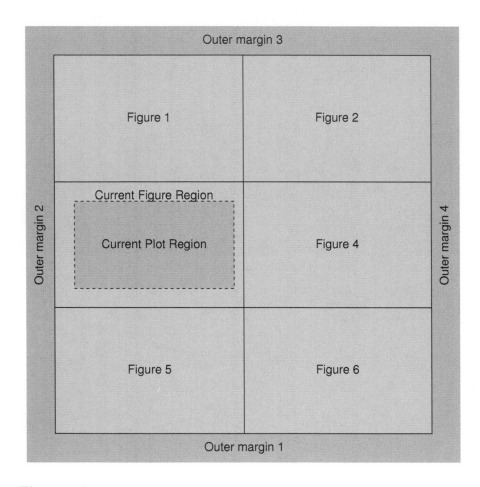

Figure 3.2
Multiple figure regions in base graphics — the outer margins, *current* figure region, and *current* plot region — when there are multiple plots on the page.

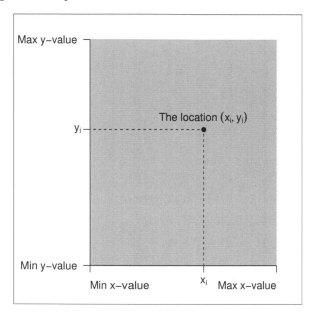

Figure 3.3
The user coordinate system in the plot region. Locations within this coordinate system are relative to the scales on the plot axes.

to this coordinate system. All of these figure margin coordinate systems are created implicitly from the arrangement of the figure margins and the setting of the user coordinate system.

The outer margins have similar sets of coordinate systems, but locations along the boundary of the inner region can only be specified in normalized coordinates (always relative to the extent of the complete outer margin). Figure 3.5 shows two of the four outer margin coordinate systems.

Sections 3.4.2 and 3.4.4 describe functions that draw output relative to the figure margin and outer margin coordinate systems.

3.1.2 The base graphics state

The base graphics system maintains a graphics "state" for the graphics window and, when drawing occurs, this state is consulted to determine where output should be drawn, what color to use, what fonts to use, and so on.

The graphics state consists of a large number of settings. Some of these settings describe the size and placement of the plot regions and coordinate systems that were described in the previous section. Some settings describe

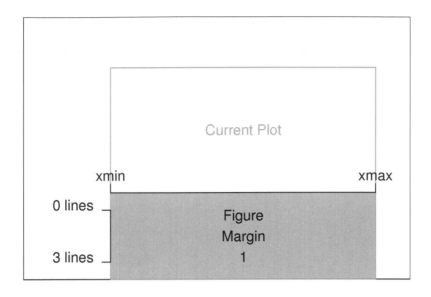

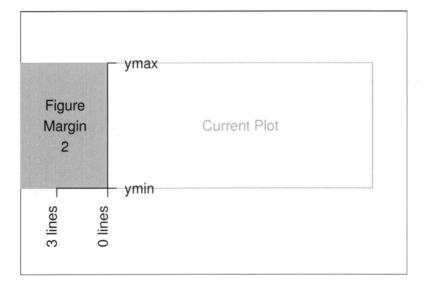

Figure 3.4

Figure margin coordinate systems. The typical coordinate systems for figure margin
1 (top plot) and figure margin 2 (bottom plot). Locations within these coordinate
systems are a combination of position along the axis scale and distance away from
the axis in multiples of lines of text.

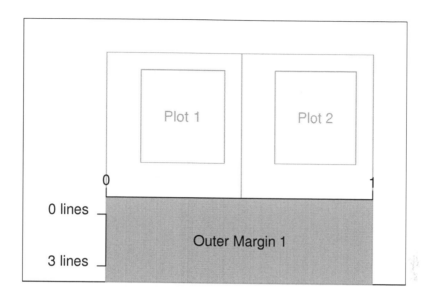

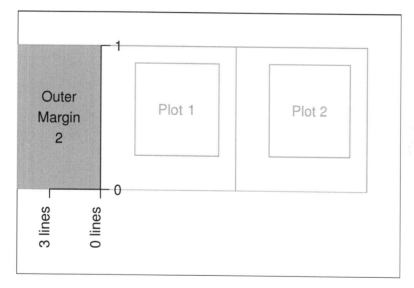

Figure 3.5

Outer margin coordinate systems. The typical coordinate systems for outer margin 1 (top plot) and outer margin 2 (bottom plot). Locations within these coordinate systems are a combination of a proportion along the inner region and distance away from the inner region in multiples of lines of text.

the general appearance of graphical output (e.g., the colors and line types that are used to draw lines and the fonts that are used to draw text) and some settings describe aspects of the output device (e.g., the physical size of the device and the current clipping region).

Tables 3.1 to 3.3 together provide a list of all of the graphics state settings and a very brief indication of their meaning. Most of the settings are described in detail in Sections 3.2 and 3.3.

The main function used to access the graphics state is the `par()` function. Simply typing `par()` will result in a complete listing of the current graphics state. A specific state setting can be queried by supplying specific setting names as arguments to `par()`. The following code queries the current state of the `col` and `lty` settings. In this case, we are asking what the current drawing color is (black) and what the current line type is (solid).

```
> par(c("col", "lty"))

$col
[1] "black"

$lty
[1] "solid"
```

The `par()` function can be used to modify base graphics state settings by specifying a value via an argument with the appropriate setting name. The following code sets new values for the `col` and `lty` settings. In this case, we are changing the drawing color to red and the line type to dashed.

```
> par(col="red", lty="dashed")
```

Modifying base graphics state settings via `par()` has a persistent effect. Settings specified in this way will hold until a different setting is specified. Settings may also be *temporarily* modified by specifying a new value in a call to a high-level graphics function such as `plot()` or a low-level graphics function such as `lines()`. The following code demonstrates this idea. First of all, the line type is permanently set to dashed using `par()`, then a plot is drawn and the lines drawn between data points in this plot are dashed. Next, a plot is drawn with a temporary line type setting of `lty="solid"` and the lines in this plot are solid. When the third plot is drawn, the permanent line type setting of `lty="dashed"` is back in effect so the lines are again dashed (see Figure 3.6).*

---

*The data used in this example are daily closing prices of major European stock indices available as the `EuStockMarkets` object from the **datasets** package.

Table 3.1

High-level base graphics state settings. This set of graphics state settings can be queried and set via the `par()` function *and* can be used as arguments to other graphics functions (e.g., `plot()` or `lines()`). Each setting is described in more detail in the relevant **Section**.

Setting	Description	Section
`adj`	Justification of text	3.2.3
`ann`	Draw plot labels and titles?	3.2.3
`bg`	Background color	3.2.1
`bty`	Type of box drawn by `box()`	3.2.5
`cex`	Size of text (multiplier)	3.2.3
also `cex.axis`, `cex.lab`, `cex.main`, `cex.sub`		
`col`	Color of lines and data symbols	3.2.1
also `col.axis`, `col.lab`, `col.main`, `col.sub`		
`family`	Font family for text	3.2.3
`fg`	Foreground color	3.2.1
`font`	Font face (bold, italic) for text	3.2.3
also `font.axis`, `font.lab`, `font.main`, `font.sub`		
`lab`	Number of ticks on axes	3.2.5
`las`	Rotation of text in margins	3.2.3
`lend`	Line end/join style	3.2.2
also `ljoin`, `lmitre`		
`lty`	Line type (solid, dashed)	3.2.2
`lwd`	Line width	3.2.2
`mgp`	Placement of axis ticks and tick labels	3.2.5
`pch`	Data symbol type	3.2.4
`srt`	Rotation of text in plot region	3.2.3
`tck`	Length of axis ticks (relative to plot size)	3.2.5
`tcl`	Length of axis ticks (relative to text size)	3.2.5
`xaxp`	Number of ticks on x-axis	3.2.5
`xaxs`	Calculation of scale range on x-axis	3.2.5
`xaxt`	X-axis style (standard, none)	3.2.5
`xpd`	Clipping region	3.2.7
`yaxp`	Number of ticks on y-axis	3.2.5
`yaxs`	Calculation of scale range on y-axis	3.2.5
`yaxt`	Y-axis style (standard, none)	3.2.5

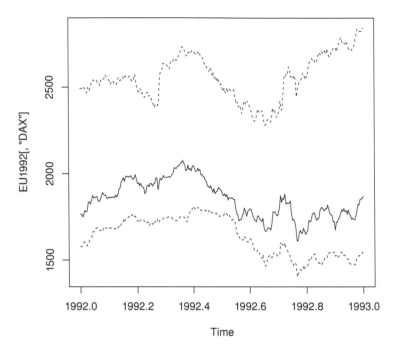

Figure 3.6
Persistent versus temporary graphical settings. The line type was set permanently
to dashed with the `par()` function to draw a plot containing the top line, then the
line type was temporarily set to solid in a call to `lines()` for the middle line, then
the line type reverted to the permanent dashed setting for the call to `lines()` for
the bottom line.

```
> EU1992 <- window(EuStockMarkets, 1992, 1993)
> par(lty="dashed")
> plot(EU1992[,"DAX"], ylim=range(EU1992))
> lines(EU1992[,"CAC"], lty="solid")
> lines(EU1992[,"FTSE"])
```

Only some of the graphics state settings can be set temporarily in calls to
graphics functions. For example, the `mfrow` setting may not be set in this way
and can only be set using `par()`. These "low-level" settings are listed in Table
3.2.

Table 3.2

Low-level base graphics state settings. This set of graphics state settings can only be queried and set via the **par**() function. Each setting is described in more detail in the relevant **Section**.

Setting	Description	Section
fig	Location of figure region (normalized)	3.2.6
fin	Size of figure region (inches)	3.2.6
lheight	Line spacing (multiplier)	3.2.3
mai	Size of figure margins (inches)	3.2.6
mar	Size of figure margins (lines of text)	3.2.6
mex	Line spacing in margins	3.2.6
mfcol	Number of figures on a page	3.3.1
mfg	Which figure is used next	3.3.1
mfrow	Number of figures on a page	3.3.1
new	Has a new plot been started?	3.2.8
oma	Size of outer margins (lines of text)	3.2.6
omd	Location of inner region (normalized)	3.2.6
omi	Size of outer margins (inches)	3.2.6
pin	Size of plot region (inches)	3.2.6
plt	Location of plot region (normalized)	3.2.6
ps	Size of text (points)	3.2.3
pty	Aspect ratio of plot region	3.2.6
usr	Range of scales on axes	3.4.5
xlog	Logarithmic scale on x-axis?	3.2.5
ylog	Logarithmic scale on y-axis?	3.2.5

Table 3.3
Read-only base graphics state settings. This set of graphics state settings can only be queried (via the `par()` function). Each setting is described in more detail in the relevant **Section**.

Setting	Description	Section
cin	Size of a character (inches)	3.4.5
cra	Size of a character ("pixels")	3.4.5
cxy	Size of a character (user coordinates)	3.4.5
din	Size of graphics device (inches)	3.4.5
page	Will the next plot start a new page?	3.3.1

A small set of graphics state settings cannot be modified at all and can only be queried using `par()`. For example, there is no function to allow the user to modify the size of the current device (after the device has been created), but its size (in inches) may be obtained using `par("din")`. These "read-only" settings are listed in Table 3.3.

It is possible to have more than one graphics window open at the same time (see Section 9.1). Every graphics window has its own graphics state and calls to `par()` only affect the base graphics state of the currently active graphics window (see Section 9.1).

3.2   Controlling the appearance of plots

This section is concerned with the appearance of plots, which means the colors, line types, fonts and so on that are used to draw a plot. As described in Section 3.1.2, these features are controlled via base graphics state settings and values are specified for the settings either with a call to the `par()` function or as arguments to a specific graphics function such as `plot()`. For example, there is a setting called `col` to control the color of output (see Section 3.2.1). This can be set permanently using `par()` with an expression of the form:

```
par(col="red")
```

This will affect all subsequent graphical output. Alternatively, the setting can be specified as an argument to a high-level function using an expression of the form:

```
plot(..., col="red")
```

This will affect the output just for that plot. Finally, the setting can be used as an argument to a low-level function, as in the expression below.

```
lines(..., col="red")
```

This demonstrates that the setting can be used to control the appearance of just a single piece of graphical output.

There are many individual settings that affect the appearance of a plot, but they can be grouped in terms of what aspects of a plot the settings affect. Each of the following sections details a particular group of settings, including a description of the role of individual settings. There are sections on specifying colors; how to control the appearance of lines, text, data symbols, and axes; how to control the size and location of the various plotting regions; clipping (only drawing output on certain parts of the page); and specifying what should happen when a high-level function is called to start a new plot.

The appearance of plots is also affected by the location and size of the plotting regions, but this is dealt with separately in Section 3.3.

The following sections provide some simple examples of how to specify the settings for the base graphical parameters, but much more detail is provided in Chapter 10.

3.2.1   Colors

There are three main color settings in the base graphics state: `col`, `fg`, and `bg`.

The `col` setting is the most commonly used. The primary use is to specify the color of data symbols, lines, text, and so on that are drawn in the plot region. Unfortunately, when specified via a graphics function, the effect can vary. For example, a standard scatterplot produced by the `plot()` function will use `col` for coloring data symbols and lines, but the `barplot()` function will use `col` for filling the contents of its bars. In the `rect()` function (see Section 3.4), the `col` argument provides the color to fill the rectangle and there is a `border` argument specific to `rect()` that gives the color to draw the border of the rectangle. The effect of `col` on graphical output drawn in the margins also varies. It does not affect the color of axes and axis labels, but it does affect the output from the `mtext()` function. There are specific settings for affecting axes, labels, titles, and subtitles called `col.axis`, `col.lab`, `col.main`, and `col.sub`.

The `fg` setting is primarily intended for specifying the color of axes and borders on plots. There is some overlap between this and the specific `col.axis`, `col.main`, etc. settings described above.

The `bg` setting is primarily intended to specify the color of the background for base graphics output. This color is used to fill the entire page. As with the `col` setting, when `bg` is specified in a graphics function it can have a quite different meaning. For example, the `plot()` and `points()` functions use `bg` to specify the color for the interior of the data symbols, which can have different colors on the border (`pch` values 21 to 25; see Section 3.2.4).

Colors may be specified in a number of different ways. The most simple is to use a color name, such as `"red"` and `"blue"`, but there are many alternatives, including generating sets of colors by calling a function. Section 10.1 describes the specification of colors in R in complete detail.

Fill patterns

In some cases (e.g., when printing in black and white), it is difficult to make use of different colors to distinguish between different elements of a plot. Using different levels of gray can be effective, but another option is to make use of some sort of fill pattern, such as cross-hatching. These should be used with caution because it is very easy to create visual effects that are distracting.

In base graphics, there is only limited support for fill patterns and they can only be applied to rectangles and polygons. It is possible to fill a rectangle or polygon with a set of lines drawn at a certain angle, with a specific separation between the lines. A `density` argument controls the separation between the lines (in terms of lines per inch) and an `angle` argument controls the angle of the lines (in terms of degrees anti-clockwise from 3 o'clock). Examples of the use of fill patterns are given in Figures 2.8, 3.20, and their associated code.

These settings can only be controlled via arguments to the functions `rect()`, `polygon()`, `hist()`, `barplot()`, `pie()`, and `legend()` (and *not* via `par()`).

3.2.2   Lines

There are five graphics state settings for controlling the appearance of lines. The `lty` setting describes the type of line to draw (e.g., solid, dashed, or dotted), the `lwd` setting describes the width of lines, and the `ljoin`, `lend`, and `lmitre` settings control how the ends and corners in lines are drawn (rounded or pointy).

The line type can be specified as a character value, for example, `"solid"`, `"dashed"`, or `"dotted"`, and the line width is given as a number, where 1 corresponds to 1/96 inch (which is roughly 1 pixel on many computer screens).

The scope of these settings again differs depending on the graphics function being called. For example, for standard scatterplots, the setting only applies to lines drawn within the plot region. In order to affect the lines drawn as part of the axes, the `lty` setting must be passed directly to the `axis()` function (see Section 3.4.4).

Section 10.2 describes the specification of line styles in R in complete detail.

3.2.3 Text

There are a large number of base graphics state settings for controlling the appearance of text. The size of text is controlled via `ps` and `cex`; the font is controlled via `font` and `family`; the justification of text is controlled via `adj`; and the angle of rotation is controlled via `srt`.

There is also an `ann` setting, which indicates whether titles and axis labels should be drawn on a plot. This is intended to apply to high-level functions, but is not guaranteed to work with all such functions (especially functions from extension packages). There are examples of the use of `ann` as an argument to high-level plotting functions in Section 3.4.1.

Text size

The size of text is ultimately a numerical value specifying the size of the font in "points." The font size is controlled by two settings: `ps` specifies an absolute font size setting (e.g., `ps=9`), and `cex` specifies a multiplicative modifier (e.g., `cex=1.5`). The final font size specification is simply `fontsize * cex`.

As with specifying color, the scope of a `cex` setting can vary depending on where it is given. When `cex` is specified via `par()`, it affects most text. However, when `cex` is specified via `plot()`, it only affects the size of data symbols. There are special settings for controlling the size of text that is drawn as axis tick labels (`cex.axis`), text that is drawn as axis labels (`cex.lab`), text in the title (`cex.main`), and text in the subtitle (`cex.sub`).

Specifying fonts

The font used for drawing text is controlled by the settings `family` and `font`.

The `family` setting is a character value giving the name of a specific font family, such as `"Times Roman"`, or a generic family style, such as `"serif"`, `"sans"` (sans-serif), or `"mono"` (monospaced). Specific font families will only

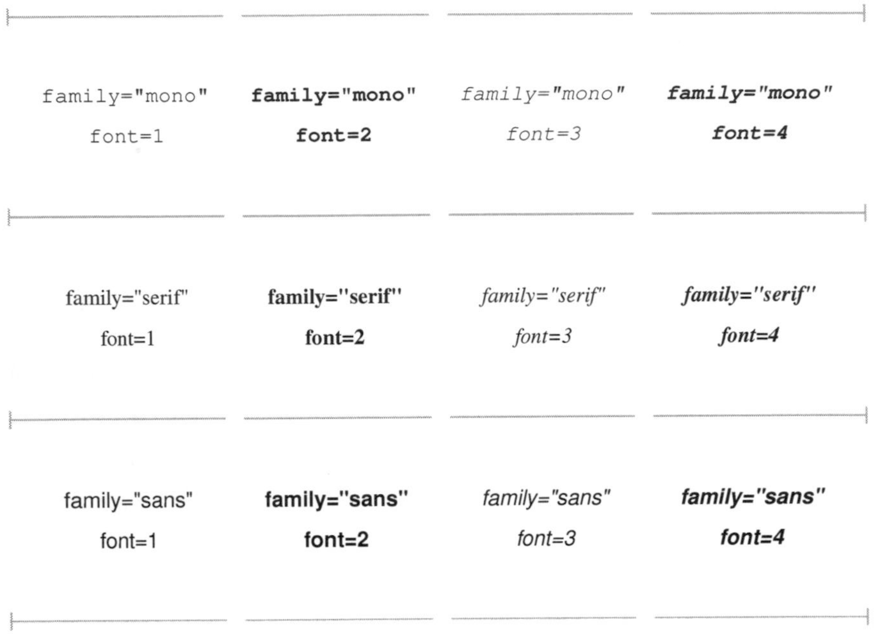

Figure 3.7
Font families and font faces. The appearance of the twelve font family and font face combinations that are available in the base graphics system.

be available if they are installed on the operating system that R is run on, but the generic family styles are always available.

The `font` setting is a numeric value that selects between normal text (1), **bold** (2), *italic* (3), and ***bold-italic*** (4). Similar to color and text size, the `font` setting applies mostly to text drawn in the plot region. There are additional settings specifically for labels (`font.lab`), and titles (`font.main` and `font.sub`). Figure 3.7 demonstrates the 12 basic font family and face combinations.

The specification of fonts in R is described in great detail in Section 10.4.

Justification of text

The `adj` setting is a value from 0 to 1 indicating the horizontal justification of text strings (0 means left-justified, 1 means right-justified and a value of 0.5 centers text).

The meaning of the `adj` setting depends on whether text is being drawn in the plot region, in the figure margins, or in the outer margins. In the plot region, the justification is relative to the (x, y) location at which the text is being drawn. In this context, it is also possible to specify two values for the setting and the second value is taken as a vertical justification for the text. Furthermore, non-finite values (`NA`, `NaN`, or `Inf`) may be specified for the justification and this is taken to mean "exact" centering (see below).

There is only a difference between a justification value of 0.5 and a non-finite justification value for vertical justification. In this case, a setting of 0.5 means text is vertically centered based on the height of the text above the text baseline (i.e., ignoring "descenders" like the tail on a "y"). A non-finite value means that text is vertically centered based on the full height of the text (including descenders). Figure 3.8 shows how various `adj` settings affect the alignment of text in the plot region.

In the figure margins and outer margins, the meaning of the `adj` setting depends on the `las` setting (see below). When margin text is parallel to the axis, `adj` specifies *both* the location and the justification of the text. For example, a value of 0 means that the text is left-justified *and* that the text is located at the left end of the margin. When text is perpendicular to the axis, the `adj` setting only affects justification. Furthermore, the `adj` setting only affects "horizontal" justification (justification in the reading direction) for text in the margins. Section 3.4.2 contains more information about the justification of text in the plot margins.

Rotating text

The `srt` setting specifies a rotation angle anti-clockwise from the positive x-axis, in degrees. This will only affect text drawn in the plot region (text drawn by the `text()` function; see Section 3.4.1). Text can be drawn at any angle within the plot region.

In the figure and outer margins, text may only be drawn at angles that are multiples of 90°, and this angle is controlled by the `las` setting. A value of 0 means text is always drawn parallel to the relevant axis (i.e., horizontal in margins 1 and 3, and vertical in margins 2 and 4). A value of 1 means text is always horizontal, 2 means text is always perpendicular to the relevant axis, and 3 means text is always vertical.

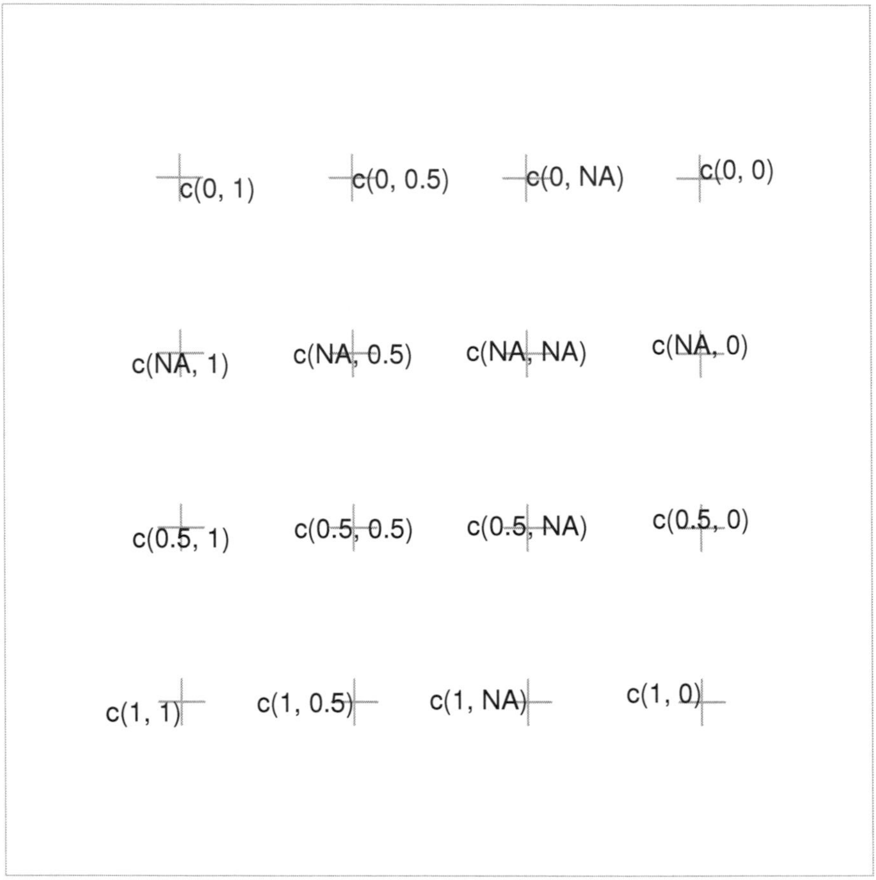

Figure 3.8
Alignment of text in the plot region. The `adj` graphical setting may be given two
values, `c(`*hjust*, *vjust*`)`, where *hjust* specifies horizontal justification and *vjust* specifies vertical justification. Each piece of text in the diagram is justified relative to a
gray cross to represent the effect of the relevant `adj` setting. The vertical adjustment
for `NA` is subtly different from the vertical adjustment for `0.5`.

Figure 3.9
The first six data symbols that are available in base graphics. In the diagram, the relevant integer value for the **pch** setting is shown in gray to the left of the corresponding symbol.

Multi-line text

The spacing between multiple lines of text is controlled by the **lheight** setting, which is a multiplier applied to the natural height of a line of text. For example, **lheight=2** specifies double-spaced text. This setting can only be specified via **par()**.

3.2.4   Data symbols

The data symbol used for plotting points is controlled by the **pch** setting. This can be an integer value to select one of a fixed set of data symbols, or a single character. For example, specifying **pch=0** produces an open square, **pch=1** produces an open circle, and **pch=2** produces an open triangle (see Figure 3.9). Specifying **pch="#"** means that a hash character will be plotted at each data location.

Some of the predefined data symbols (**pch** between 21 and 25) allow a fill color separate from the border color, with the **bg** setting controlling the fill color in these cases.

Section 10.3 describes the possible set of data symbols in more detail.

The size of the data symbols is linked to the size of text and is affected by the **cex** setting. If the data symbol is a character, the size will also be affected by the **ps** setting.

The **type** setting controls how data are represented in a plot. A value of "p" means that data symbols are drawn at each (x, y) location. The value "l" means that the (x, y) locations are connected by lines. A value of "b" means that both data symbols and lines are drawn. The **type** setting may also have the value "o", which means that data symbols are "over-plotted" on lines (with the value "b", the lines stop short of each data symbol). It is also possible to specify the value "h", which means that vertical lines are drawn from the x-axis to the (x, y) locations (the appearance is like a barplot with very thin bars). Two further values, "s" and "S" mean that (x, y) locations are joined in a city-block fashion with lines going horizontally then vertically (or vertically then horizontally) between each data location. Finally, the value "n" means that nothing is drawn at all.

Figure 3.10 shows simple examples of the different plot types. This setting is most often specified within a call to a high-level function (e.g., plot()) rather than via par().

3.2.5 Axes

By default, the base graphics system produces axes with sensible labels and tick marks at sensible locations. If the axis does not look right, there are a number of graphical state settings specifically for controlling aspects such as the number of tick marks and the positioning of labels. These are described below. If none of these gives the desired result, the user may have to resort to drawing the axis explicitly using the axis() function (see Section 3.4.4).

The lab setting in the base graphics state is used to control the number of tick marks on the axes. The setting is only used as a starting point for the algorithm R uses to determine sensible tick locations so the final number of tick marks that are drawn could easily differ from this specification. The setting takes two values: the first specifies the number of tick marks on the x-axis and the second specifies the number of tick marks on the y-axis.

The xaxp and yaxp settings also relate to the number and location of the tick marks on the axes of a plot. This setting is almost always calculated by R for each new plot so user settings are usually overridden (see Section 3.4.4 for an exception to this rule). In other words, it only makes sense to query this setting for its current value. The settings consist of three values: the first two specify the location of the left-most and right-most tick marks (bottom and top tick marks for the y-axis), and the third value specifies how many intervals there are between tick marks. When a log transformation is in effect for an axis, the three values have a different and much more complicated meaning altogether (see the on-line help page for par()).

The mgp setting controls the distance that the components of the axes are drawn away from the edge of the plot region. There are three values representing the positioning of the overall axis label, the tick mark labels, and the lines for the ticks. The values are in terms of lines of text away from the edges of the plot region. The default value is c(3, 1, 0). Figure 3.11 gives an example of different mgp settings.

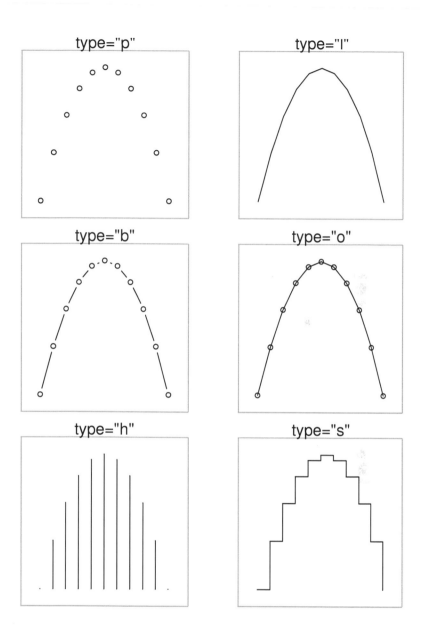

Figure 3.10

Basic plot types. Plotting the same data with different plot `type` settings. In each case, the output is produced by an expression of the form `plot(x, y, type=`*something*`)`, where the relevant value of `type` is shown above each plot.

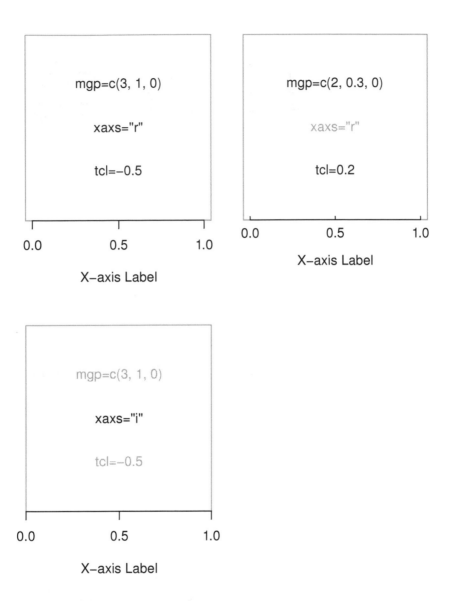

Figure 3.11
Different axis styles. The top-left plot demonstrates the default axis settings for an
x-axis. The top-right plot shows the effects of specifying different positions for the
axis labels (the tick labels and axis labels are closer to the plot region) and different
lengths for the tick marks and the bottom-left plot shows the effect of specifying an
"internal" axis range calculation.

The tck and tcl settings control the length of tick marks. The tcl setting specifies the length of tick marks as a fraction of the height of a line of text. The sign dictates the direction of the tick marks — a negative value draws tick marks outside the plot region and a positive value draws tick marks inside the plot region. The tck setting specifies tick mark lengths as a fraction of the smaller of the physical width or height of the plotting region, but it is only used if its value is not NA (and it is NA by default). Figure 3.11 gives an example of different tcl settings.

The xaxs and yaxs settings control the "style" of the axes of a plot. By default, the setting is "r", which means that R calculates the range of values on the axis to be wider than the range of the data being plotted (so that data symbols do not collide with the boundaries of the plot region). It is possible to make the range of values on the axis exactly match the range of values in the data, by specifying the value "i". This can be useful if the range of values on the axes are being explicitly controlled via xlim or ylim arguments to a function. Figure 3.11 gives an example of different xaxs settings.

The xaxt and yaxt settings control the "type" of axes. The default value, "s", means that the axis is drawn. Specifying a value of "n" means that the axis is not drawn.

The xlog and ylog settings control the transformation of values on the axes. The default value is FALSE, which means that the axes are linear and values are not transformed. If this value is TRUE then a logarithmic transformation is applied to any values on the relevant dimension in the plot region. This also affects the calculation of tick mark locations on the axes.

When data of a special nature are being plotted (e.g., time series data), some of these settings may not apply (and may not have any sensible interpretation).

The bty setting is not strictly to do with axes, but it controls the output of the box() function, which is most commonly used in conjunction with drawing axes. This function draws a bounding box around the edges of the plot region (by default). The bty setting controls the type of box that the box() function draws. The value can be "n", which means that no box is drawn, or it can be one of "o", "l", "7", "c", "u", or "]", which means that the box drawn resembles the corresponding uppercase character. For example, bty="c" means that the bottom, left, and top borders will be drawn, but the right border will not be drawn.

In addition to these graphics state settings, many high-level plotting functions, e.g., plot(), provide arguments xlim and ylim to control the range of the scale on the axes. Section 2.6.1 has an example.

3.2.6   Plotting regions

As described in Section 3.1.1, the base graphics system defines several different regions on the graphics device. This section describes how to control the size and layout of these regions using graphics state settings. Figure 3.12 shows a diagram of some of the settings that affect the widths and horizontal placement of the regions.

The size of each margin can be controlled independently, but R will check whether an overall specification is consistent. For example, if the margins are made too big, so that there is not room left on the page for the plot region, then R will give an error message like the following:

```
Error in plot.new() : figure margins too large
```

Outer margins

By default, there are no outer margins on a page. Outer margins can be specified using the `oma` graphics state setting. This consists of four values for the four margins in the order (`bottom, left, top, right`) and values are interpreted as lines of text (a value of 1 provides space for one line of text in the margin). The margins can also be specified in inches using `omi` or in normalized device coordinates (i.e., as a proportion of the device region) using `omd`. If `omd` is used, the margins are specified in the order (`left, right, bottom, top`).

Figure regions

By default, the figure region is calculated from the settings for the outer margins and the number of figures on the page. The figure region can be specified explicitly instead, using either the `fig` setting or the `fin` state setting. The `fig` setting specifies the location, (`left, right, bottom, top`), of the figure region where each value is a proportion of the "inner" region (the page less the outer margins). The `fin` setting specifies the size, (`width, height`), of the figure region in inches and the resulting figure region is centered within the inner region.

Figure margins

The figure margins can be controlled using the `mar` state setting. This consists of four values for the four margins in the order (`bottom, left, top, right`) where each value represents a number of lines of text. The default values are

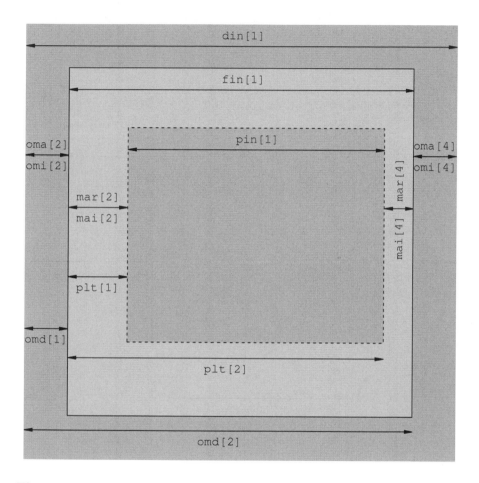

Figure 3.12

Graphics state settings controlling plot regions. These are some of the settings that control the widths and horizontal locations of the plot regions. For ease of comparison, this diagram has the same layout as Figure 3.1: the central gray rectangle represents the plot region, the lighter gray rectangle around that is the figure region, and the darker gray rectangle around that is the outer margins. A similar diagram could be produced for settings controlling heights and vertical locations.

`c(5, 4, 4, 2) + 0.1`. The margins may also be specified in terms of inches using `mai`.

The `mex` setting controls the size of a "line" in the margins. This does not affect the size of text drawn in the margins, but is used to multiply the size of text to determine the height of one line of text in the margins.

Plot regions

By default, the plot region is calculated from the figure region less the figure margins. The location and size of the plot region may be controlled explicitly instead, using the `plt`, `pin`, or `pty` settings. The `plt` setting allows the user to specify the location of the plot region, `(left, right, bottom, top)`, where each value is a proportion of current figure region. The `pin` setting specifies the size of the plot region, `(width, height)`, in terms of inches.

The `pty` setting controls how much of the available space (figure region less figure margins) the plot region occupies. The default value is `"m"`, which means that the plot region occupies all of the available space. A value of `"s"` means that the plot region will take up as much of the available space as possible, but it must be "square" (i.e., its physical width will be the same as its physical height).

3.2.7   Clipping

Base graphics output is usually clipped to the plot region. This means that any output that would appear outside the plot region is not drawn. For example, in the default behavior, data symbols for `(x, y)` locations which lie outside the ranges of the axes are not drawn. Base graphics functions that draw in the margins clip output to the current figure region or to the device. Section 3.4 has information about which functions draw in which regions.

It can be useful to override the default clipping region. For example, this is necessary to draw a legend outside the plot region using the `legend()` function.

The base clipping region is controlled via the `xpd` setting. Clipping can occur either to the whole device (an `xpd` value of `NA`), to the current figure region (a value of `TRUE`), or to the current plot region (a value of `FALSE`, which is the default).

There is also a `clip()` function for setting the clipping region to be *smaller* than the plot region.

3.2.8 Moving to a new plot

As described in Section 2.1, high-level graphics functions usually start a new plot.

The `devAskNewPage()` function can be used to control whether the user is prompted before the graphics system starts a new page of output.

The graphics state includes a setting called `new`, which controls whether a function that starts a new plot will move on to the next figure region (possibly a new page). Every plot sets the value to `FALSE` so that the next plot will move on by default, but if this setting has the value `TRUE` then a new plot does not move on to the next figure region. This can be used to overlay several plots on the same figure (Section 3.4.5 has an example).

3.3 Arranging multiple plots

There are a number of ways to produce multiple plots on a single page.

The number of plots on a page, and their placement on the page, can be controlled directly by specifying the base graphics state settings `mfrow` or `mfcol` using the `par()` function, or through a higher-level interface provided by the `layout()` function. The `split.screen()` function provides yet another approach, where a figure region can itself be treated as a complete page to split into further figure and plot regions.

These three approaches are mutually incompatible. For example, a call to the `layout()` function will override any previous `mfrow` and `mfcol` settings. Also, some high-level functions (e.g., `coplot()`) call `layout()` or `par()` themselves to create a plot arrangement, which means that the output from such functions cannot be arranged with other plots on a page (see Section 3.4.6 for further discussion; Section 12.2 describes one way to work around this limitation).

3.3.1 Using the base graphics state

The number of figure regions on a page can be controlled via the `mfrow` and `mfcol` graphics state settings. Both of these consist of two values indicating a number of rows, *nr*, and a number of columns, *nc*; these settings result in $nr \times nc$ figure regions of equal size.

The top-left figure region is used first. If the setting is made via `mfrow` then

the figure regions along the top row are used next from left to right, until that
row is full. After that, figure regions are used in the next row down, from
left to right, and so on. When all rows are full, a new page is started. For
example, the following code creates six figure regions on the page, arranged
in three rows and two columns and the regions are used in the order shown
in Figure 3.13(a).

```
> par(mfrow=c(3, 2))
```

If the setting is made via `mfcol`, figure regions are used in a column-first order
instead of a row-first order.

The order in which figure regions are used can be controlled explicitly by using
the `mfg` setting to specify the next figure region. This setting consists of two
values that indicate the row and column of the next figure to use.

The read-only `page` setting can be queried to determine whether the next
high-level graphics function is going to start a new page.

3.3.2 Layouts

The `layout()` function provides an alternative to the `mfrow` and `mfcol` settings. The primary difference is that the `layout()` function allows the creation
of multiple figure regions of *unequal* size.

The simple idea underlying the `layout()` function is that it divides the inner
region of the page into a number of rows and columns, but the heights of rows
and the widths of columns can be independently controlled, *and* a figure can
occupy more than one row or more than one column.

The first argument (and the only required argument) to the `layout()` function
is a matrix. The number of rows and columns in the matrix determines the
number of rows and columns in the layout.

The contents of the matrix are integer values that determine which rows and
columns each figure will occupy. The following layout specification is identical
to `par(mfrow=c(3, 2))`.

```
> layout(matrix(c(1, 2, 3, 4, 5, 6), byrow=TRUE, ncol=2))
```

It may be easier to imagine the arrangement of figure regions if the matrix
is specified using `cbind()` or `rbind()`. The code below repeats the previous
example, but uses `rbind()` to specify the layout matrix.

```
> layout(rbind(c(1, 2),
               c(3, 4),
               c(5, 6)))
```

The function `layout.show()` may be helpful for visualizing the figure regions that are created. The following code creates a figure visualizing the layout created in the previous example (see Figure 3.13(a)).

```
> layout.show(6)
```

The contents of the layout matrix determine the order in which the resulting figure regions will be used. The following code creates a layout with exactly the same rows and columns as the previous one, but the figure regions will be used in the reverse order (see Figure 3.13(b)).

```
> layout(rbind(c(6, 5),
               c(4, 3),
               c(2, 1)))
```

By default, all row heights are the same and all column widths are the same size and the available inner region is divided up equally. The `heights` arguments can be used to specify that certain rows are given a greater portion of the available height (for all of what follows, the `widths` argument works analogously for column widths). When the available height is divided up, the proportion of the available height given to each row is determined by dividing the row heights by the sum of the row heights. For example, in the following layout there are two rows and one column. The top row is given two thirds of the available height, $2/(2+1)$, and the bottom row is given one third, $1/(2+1)$. Figure 3.13(c) shows the resulting layout.

```
> layout(matrix(c(1, 2)), heights=c(2, 1))
```

In the examples so far, the division of row heights has been completely independent of the division of column widths. The widths and heights can be forced to correspond as well so that, for example, a height of 1 corresponds to the same physical distance as a width of 1. This allows control over the aspect ratio of the resulting figure. The `respect` argument is used to force this correspondence. The following code is the same as the previous example except that the `respect` argument is set to `TRUE` (see Figure 3.13(d)).

```
> layout(matrix(c(1, 2)), heights=c(2, 1),
         respect=TRUE)
```

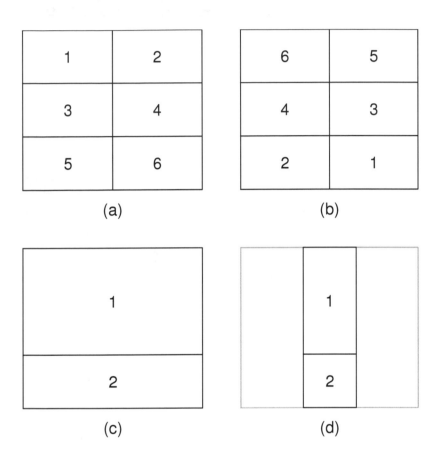

Figure 3.13
Some basic layouts: (a) A layout that is identical to `par(mfrow=c(3, 2))`; (b) Same as (a) except the figures are used in the reverse order; (c) A layout with unequal row heights; (d) Same as (c) except the layout widths and heights "respect" each other.

It is also possible to specify heights of rows and widths of columns in absolute terms. The `lcm()` function can be used to specify heights and widths for a layout in terms of centimeters. The following code is the same as the previous example, except that a third, empty region is created to provide a vertical gap of 0.5 cm between the two figures (see Figure 3.14(a)). The 0 in the first matrix argument means that no figure occupies that region.

```
> layout(matrix(c(1, 0, 2)),
          heights=c(2, lcm(0.5), 1),
          respect=TRUE)
```

This next piece of code demonstrates that a figure may occupy more than one row or column in the layout. This extends the previous example by adding a second column and creating a figure region that occupies both columns of the bottom row. In the matrix argument, the value 2 appears in both columns of row 3 (see Figure 3.14(b)).

```
> layout(rbind(c(1, 3),
               c(0, 0),
               c(2, 2)),
          heights=c(2, lcm(0.5), 1),
          respect=TRUE)
```

Finally, it is possible to specify that only certain rows and columns should respect each other's heights/widths. This is done by specifying a matrix for the `respect` argument. In the following code, the previous example is modified by specifying that only the first column and the last row should respect each other's widths/heights. In this case, the effect is to ensure that the width of figure region 1 is the same as the height of figure region 2, but the width of figure region 3 is free to expand to the available width (see Figure 3.14(c)).

```
> layout(rbind(c(1, 3),
               c(0, 0),
               c(2, 2)),
          heights=c(2, lcm(0.5), 1),
          respect=rbind(c(0, 0),
                        c(0, 0),
                        c(1, 0)))
```

3.3.3 The split-screen approach

The `split.screen()` function provides yet another way to divide the page into a number of figure regions. The first argument, **figs**, is either two

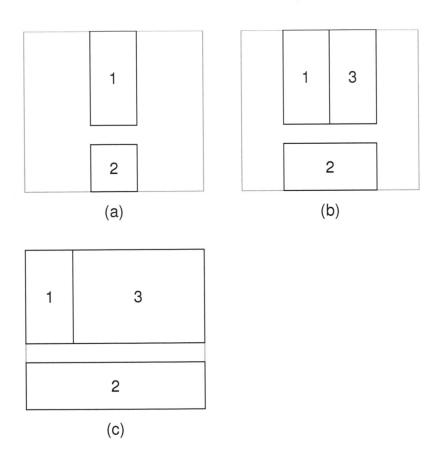

Figure 3.14
Some more complex layouts: (a) A layout with a row height specified in centimeters;
(b) A layout with a figure occupying more than one column; (c) Same as (b), but
with only column 1 and row 3 respected.

values specifying a number of rows and columns of figures (i.e., like the `mfrow` setting), or a matrix containing a figure region location, (`left`, `right`, `bottom`, `top`), on each row (i.e., like a `par(fig)` setting on each row).

Having established figure regions in this manner, a figure region is used by calling the `screen()` function to select a region. This means that the order in which figures are used is completely under the user's control, and it is possible to reuse a figure region, though there are dangers in doing so (the on-line help for `split.screen()` provides further discussion). The function `erase.screen()` can be used to clear a defined screen and `close.screen()` can be used to remove one or more screen definitions.

An even more useful feature of this approach is that each figure region can itself be divided up by a further call to `split.screen()`. This allows complex arrangements of plots to be created.

The downside to this approach is that it does not fit very nicely with the underlying base graphics system model (see Section 3.1). The recommended way to achieve complex arrangements of plots is via the `layout()` function from the previous section or by using the **grid** graphics system (see Part II), possibly in combination with base graphics high-level functions (see Chapter 12).

3.4 Annotating plots

Sometimes it is not enough to be able to modify the default output from high-level functions and further graphical output must be added, using low-level functions, to achieve the desired result (see, for example, Figure 1.3). R graphics in general is fundamentally oriented to supporting the annotation of plots — the ability to add graphical output to an existing plot. In particular, the regions and coordinate systems used in the construction of a plot remain available for adding further output to the plot. For example, it is possible to position a text label relative to the scales on the axes of a plot.

3.4.1 Annotating the plot region

Most low-level graphics functions that add output to an existing plot, add the output to the plot region. In other words, locations are specified relative to the user coordinate system (see Section 3.1.1).

Table 3.4
The low-level base graphics functions for drawing basic graphical primitives.

Function	Description
`points()`	Draw data symbols at locations (x, y)
`lines()`	Draw lines between locations (x, y)
`segments()`	Draw line segments between (x0, y0) and (x1, y1)
`arrows()`	Draw line segments with arrowheads at the end(s)
`xspline()`	Draw a smooth curve relative to control points (x, y)
`rect()`	Draw rectangles with bottom-left corner at (xl, yb) and top-right corner at (xr, yt)
`polygon()`	Draw one or more polygons with vertices (x, y)
`polypath()`	Draw a single polygon made up of one or more paths with vertices (x, y)
`rasterImage()`	Draw a bitmap image
`text()`	Draw text at locations (x, y)

Graphical primitives

This section describes the graphics functions that provide the most basic graphics output (lines, rectangles, text, etc). Table 3.4 provides a complete list.

The most common use of this facility is to add extra sets of data to a plot. The `lines()` function draws lines between (x, y) locations, and the `points()` function draws data symbols at (x, y) locations. The following code demonstrates a common situation where three different sets of y-values, recorded at the same set of x-values, are plotted together on the same plot (see the left-hand plot in Figure 3.15).

First, we extract just a few days of data from the `EuStockMarkets` time series and plot the closing price from one market as a gray line (`type="l"` and `col="gray"`). The scale on the y-axis is set, using `ylim`, to ensure that there will be room on the plot for all of the data series.

```
> EUdays <- window(EuStockMarkets, c(1992,1), c(1992,10))
> plot(EUdays[,"DAX"], ylim=range(EUdays), ann=FALSE,
        axes=FALSE, type="l", col="gray")
```

Now a set of points are added for the first set of closing prices, then lines and points are added for the closing prices of two other markets.

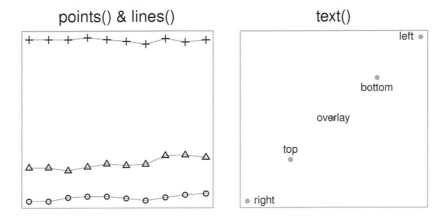

Figure 3.15
Annotating the plot region of a base graphics plot. The left-hand plot shows points
and extra lines being added to an initial line plot. The right-hand plot shows text
being added to an initial scatterplot.

```
> points(EUdays[,"DAX"])
> lines(EUdays[,"CAC"], col="gray")
> points(EUdays[,"CAC"], pch=2)
> lines(EUdays[,"FTSE"], col="gray")
> points(EUdays[,"FTSE"], pch=3)
```

It is also possible to draw text at (x, y) locations with the text() function.
This is useful for labeling data locations, particularly using the pos argument
to offset the text so that it does not overlay the corresponding data symbols.
The following code creates a diagram demonstrating the use of text() (see
the right-hand plot in Figure 3.15). Again, some data are created and (gray)
data symbols are plotted at the (x, y) locations.

```
> x <- 1:5
> y <- x
> plot(x, y, ann=FALSE, axes=FALSE, col="gray", pch=16)
```

Now some text labels are added, with each one offset in a different way from
the (x, y) location. Notice that the arguments to text() may be vectors so
that several pieces of text are drawn by the one function call.

```
> text(x[-3], y[-3], c("right", "top", "bottom", "left"),
        pos=c(4, 3, 1, 2))
> text(3, 3, "overlay")
```

Like the `plot()` function, the `text()`, `lines()`, and `points()` functions are
generic. This means that they have flexible interfaces for specifying the data
for the (x, y) locations, or they produce different output when given objects
of a particular class in the x argument. For example, both `lines()`, and
`points()` will accept formulae for specifying the (x, y) locations and the
`lines()` function will behave sensibly when given a `ts` (time series) object to
draw.

The `text()` function normally takes a character value to draw, but it will
also accept an R expression (as produced by the `expression()` function),
which can be used to produce a mathematical formula with special symbols
and formatting. For example, the following code draws the formula $\sqrt{2\pi\sigma^2}$.
Section 10.5 describes this facility in more detail.

```
> text(0.5, 0.5, expression(sqrt(2*pi*sigma^2)))
```

As a parallel to the `matplot()` function (see Section 2.5), there are functions
`matpoints()` and `matlines()` specifically for adding lines and data symbols
to a plot, given x or y as matrices.

Having access to graphical primitives not only makes it easy to add new data
series to a plot and to add labels, but it also makes it possible to add arbitrary
drawing to a plot. In addition to lines, points, and text, there are graphical
primitives for drawing more complex shapes.

In order to demonstrate these other graphical primitives, the following code
produces a simple set of x- and y-values. These points will be plotted and
used to draw a variety of shapes (see Figure 3.16).

```
> t <- seq(60, 360, 30)
> x <- cos(t/180*pi)*t/360
> y <- sin(t/180*pi)*t/360
```

The `lines()` function draws a single line through several points. Missing
values in the (x, y) locations will create breaks in the line.

```
> lines(x, y)
```

An alternative is provided by the `segments()` function, which will draw several
different straight lines between pairs of end points. In the following code, a
straight line is drawn from (0, 0) to each of the (x, y) locations. Notice that
R's normal *recycling rule* behavior is applied to most arguments of graphics
functions.

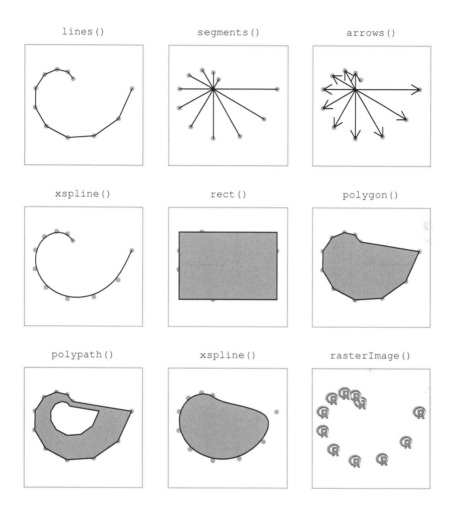

Figure 3.16

Drawing in the plot region of a base graphics plot. These pictures show some of the functions that draw more complex graphical shapes. The shapes are based on a set of (x, y) points which are drawn as light gray dots.

```
> segments(0, 0, x, y)
```

The `arrows()` function produces the same output as `segments()`, but also adds simple arrowheads at either end of the line segments. The `length` argument is used here to control the size of the arrowheads.

```
> arrows(0, 0, x[-1], y[-1], length=.1)
```

The `xspline()` function also produces a line, but the line is an X-spline, which treats the `(x, y)` locations as *control points* from which to produce a smooth curve. The smoothness of the curve is controlled by a `shape` parameter.

```
> xspline(x, y, shape=1)
```

There are also several functions for producing closed shapes. The simplest is `rect()`, which only requires a left, bottom, right, and top value to draw a rectangle (though all values can be vectors, which will result in several rectangles being drawn).

```
> rect(min(x), min(y), max(x), max(y), col="gray")
```

The `polygon()` function produces more complex shapes, using the `(x, y)` locations as vertices. Multiple polygons may be drawn using `polygon()` by inserting an `NA` value between each set of polygon vertexes. For both `rect()` and `polygon()`, the `col` argument specifies the color to *fill* the interior of the shape and the argument `border` controls the color of the line around the boundary of the shape.

```
> polygon(x, y, col="gray")
```

The `polygon()` function can draw self-intersecting polygons, but cannot represent polygons with holes. For the latter case, there is `polypath()`, which only draws a single polygon, but the polygon can be composed of more than one subpath. This allows for polygons consisting of distinct paths as well as polygons with holes.

```
> polypath(c(x, NA, .5*x), c(y, NA, .5*y),
           col="gray", rule="evenodd")
```

The `xspline()` function can also be used to create closed shapes, by specifying `open=FALSE`.

```
> xspline(x, y, shape=1, open=FALSE, col="gray")
```

Finally, there is a function, `rasterImage()`, for drawing bitmap images on a plot. The bitmap can be an external file, or it can just be a vector, matrix, or array. The following code draws the R logo at each of the (x, y) locations (code to read in the R logo is not shown; see Chapter 11 for more information).

```
> rasterImage(rlogo,
              x - .1, y - .1,
              x + .1, y + .1)
```

These examples only provide a tiny glimpse of what is possible with these graphical primitives. The possibilities are endless and a number of the examples in the remainder of this chapter provide some further demonstrations of what can be achieved by adding basic graphical shapes to a plot (see, for example, Figure 3.24).

Graphical utilities

In addition to the low-level graphical primitives of the previous section, there are a number of utility functions that provide a set of slightly more complex shapes.

The `grid()` function adds a series of grid lines to a plot. This is simply a series of line segments, but the default appearance (light gray and dotted) is suited to the purpose of providing visual cues to the viewer without interfering with the primary data symbols.

The `abline()` function provides a number of convenient ways to add a line (or lines) to a plot. The line(s) can be specified either by a slope and y-axis intercept, or as a series of x-locations for vertical lines or as a series of y-locations for horizontal lines. The function will also accept the coefficients from a linear regression analysis (even as an `"lm"` object), thereby providing a simple way to add a line of best fit to a scatterplot.

The following code annotates a basic scatterplot with a line and arrows (see the left-hand plot of Figure 3.17).

First, we plot some points in an unadorned plot.*

```
> plot(cars, ann=FALSE, axes=FALSE, col="gray", pch=16)
```

---

*The data used in this example are vehicle speeds and stopping distances that were recorded in the 1920s available as the data set `cars` in the `datasets` package.

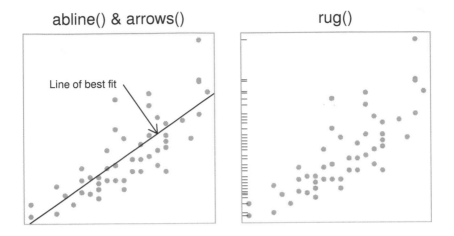

Figure 3.17
More examples of annotating the plot region of a base graphics plot. The left-hand
plot shows a line of best fit (plus a text label and arrow) being added to an initial
scatterplot. The right-hand plot shows a series of ticks being added as a rug plot on
an initial histogram.

Now a line of best fit is drawn through the data using `abline()` and a text
label and arrow are added using `text()` and `arrows()`.

```
> lmfit <- lm(dist ~ speed, cars)
> abline(lmfit)
> arrows(15, 90, 19, predict(lmfit, data.frame(speed=19)),
          length=0.1)
> text(15, 90, "Line of best fit", pos=2)
```

The `box()` function draws a rectangle around the boundary of the plot region.
The `which` argument makes it possible to draw the rectangle around the current figure region, inner region, or outer region instead. The following code
draws a gray box around the plot region in the plot above.

```
> box(col="gray")
```

The `rug()` function produces a "rug" plot along one of the axes, which consists
of a series of tick marks representing data locations. This can be useful to
represent an additional one-dimensional plot of data (e.g., in combination
with a density curve). The following code uses this function to annotate the
same scatterplot as above, with a set of tick marks on the y-axis to show the
distribution of stopping distances (see the right-hand plot of Figure 3.17).

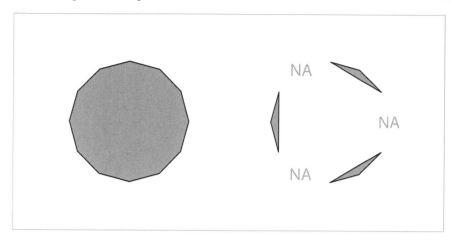

Figure 3.18
Drawing polygons using the `polygon()` function. On the left, a single polygon (dodecagon) is produced from multiple (x, y) locations. On the right, the first, fifth, and ninth values have been set to NA, which splits the output into three separate polygons. The `polygon()` function does not draw the gray NA values; those have been drawn using the `text()` function purely for the purposes of illustration.

```
> rug(cars$dist, side=2)
```

Missing values and non-finite values

R has special values representing missing observations (NA) and non-finite values (NaN and Inf). Most base graphics functions allow such values within (x, y) locations and handle them by not drawing the relevant location. For drawing data symbols or text, this means the relevant data symbol or piece of text will not be drawn. For drawing lines, this means that lines to or from the relevant location are not drawn; a gap is created in the line. For drawing rectangles, an entire rectangle will not be drawn if any of the four boundary locations is missing or non-finite.

Polygons are a slightly more complex case. For drawing polygons, a missing or non-finite value in x or y is interpreted as the end of one polygon and the start of another. Figure 3.18 shows an example. On the left, a polygon is drawn through 12 locations evenly spaced around a circle. On the right, the first, fifth, and ninth locations have been set to NA so the output is split into three separate polygons.

Missing or non-finite values can also be specified for some base graphics state settings. For example, if a color setting is missing or non-finite, then nothing is drawn (this is a brute-force way to specify a completely transparent color).

Similarly, specifying a missing value or non-finite value for `cex` means that the relevant data symbol or piece of text is not drawn.

3.4.2 Annotating the margins

There are only two functions that produce output in the figure or outer margins, relative to the margin coordinate systems (Section 3.1.1).

The `mtext()` function draws text at any location in any of the margins. The `outer` argument controls whether output goes in the figure or outer margins. The `side` argument determines which margin to draw in: 1 means the bottom margin, 2 means the left margin, 3 means the top margin, and 4 means the right margin.

Text is drawn a number of lines of text away from the edges of the plot region for figure margins or a number of lines away from the edges of the inner region for outer margins. In the figure margins, the location of the text along the margin can be specified relative to the user coordinates on the relevant axis using the `at` argument. In some cases it is possible to specify the location as a proportion of the length of the margin using the `adj` argument, but this is dependent on the value of the `las` state setting (see page 65). For certain `las` settings, the `adj` argument instead controls the justification of the text relative to a position chosen by the `las` argument. There is also a `padj` argument for controlling the "vertical" justification of text in the margins (the justification of the text perpendicular to the reading direction of the text).

The `title()` function is essentially a specialized version of `mtext()`. It is more convenient for producing a few specific types of output, but much less flexible than `mtext()`. This function can be used to produce a main title for a plot (in the top figure margin), axis labels (in the left and bottom figure margins), and a subtitle for a plot (in the bottom margin below the x-axis label). The output from this function is heavily influenced by various graphics state settings, such as `cex.main` and `col.main`, which control the size and color of the title.

Just like the `text()` function, which draws text in the plot region, the functions that draw text in the margins all accept not only a character value, but also an R expression, so that axis labels and plot titles can include special symbols and formatting (see Section 10.5).

With a little extra effort, it is also possible to produce graphical output in the figure or outer margins using the functions that normally draw in the plot region (e.g., `points()` and `lines()`). In order to do this, the clipping region of the plot must first be set using the `xpd` state setting (see Section 3.2.7). This approach is not very convenient because the functions are drawing relative to user coordinates rather than locations relative to the margin co-

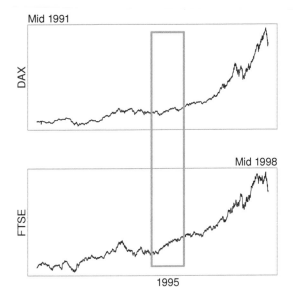

Figure 3.19
Annotating the margins of a base graphics plot. Text has been added in margin 3 of the top plot and in margins 1 and 3 in the bottom plot. Thick gray lines have been added to both plots (and overlapped so that it appears to be a single rectangle across the plots).

ordinate systems. Nevertheless, it can sometimes be useful and the functions `grconvertX()` and `grconvertY()` can help with converting locations between coordinate systems.

The following code demonstrates the use of `mtext()` and a simple application of using `lines()` outside the plot region for drawing what appears to be a rectangle extending across two plots (see Figure 3.19).*

First of all, the `mfrow` setting is used to set up an arrangement of two figure regions, one above the other. The clipping region is set to the entire device using xpd=NA.

```
> par(mfrow=c(2, 1), xpd=NA)
```

The first data set is plotted as a line on the top plot and a label is added at the left end of figure margin 3. In addition, thick gray lines are drawn to

---

*This example was motivated by a question to R-help on December 14, 2004 with subject: "drawing a rectangle through multiple plots".

represent the top of the rectangle, with the lines deliberately extending well
below the bottom of the plot. The label `"DAX"` is drawn in figure margin 2.

```
> plot(EuStockMarkets[,"DAX"], type="l", axes=FALSE,
      xlab="", ylab="", main="")
> box(col="gray")
> mtext("Mid 1991", adj=0, side=3)
> lines(x=c(1995, 1995, 1996, 1996),
        y=c(-1000, 6000, 6000, -1000),
        lwd=3, col="gray")
> mtext("DAX", side=2, line=0)
```

The second data set is plotted as a line in the bottom plot, a label is added
to this plot at the right end of figure margin 3, and another label is drawn
beneath the x-location `1995.5` in figure margin 1. Finally, thick gray lines are
drawn to represent the bottom of the rectangle, again deliberately extending
these above the plot, and the label `"FTSE"` is drawn in figure margin 2. The
thick gray lines overlap the lines drawn with respect to the top plot to create
the impression of a single rectangle traversing both plots.

```
> plot(EuStockMarkets[,"FTSE"], type="l", axes=FALSE,
      xlab="", ylab="", main="")
> box(col="gray")
> mtext("Mid 1998", adj=1, side=3)
> mtext("1995", at=1995.5, side=1)
> lines(x=c(1995, 1995, 1996, 1996),
        y=c(7000, 2500, 2500, 7000),
        lwd=3, col="gray")
> mtext("FTSE", side=2, line=0)
```

3.4.3   Legends

The base graphics system provides the `legend()` function for adding a legend
or key to a plot. The legend is usually drawn within the plot region, and
is located relative to user coordinates. The function has many arguments,
which allow for a great deal of flexibility in the specification of the contents
and layout of the legend. The following code demonstrates a couple of typical
uses.

The first example shows a scatterplot with a legend to relate group names to
different symbols (see the top plot in Figure 3.20). The first two arguments
give the position of the top-left corner of the legend, relative to the user
coordinate system. The third argument provides labels for the legend and,

because the `pch` argument is also specified, data symbols are drawn beside each label.

```
> with(iris,
       plot(Sepal.Length, Sepal.Width,
            pch=as.numeric(Species), cex=1.2))
> legend(6.1, 4.4, c("setosa", "versicolor", "virginica"),
         cex=1.5, pch=1:3)
```

The next example shows a barplot with a legend to relate group names to different fill patterns (see the bottom plot in Figure 3.20). In this example, the `angle`, `density`, and `fill` arguments are specified, so small rectangles with fill patterns are drawn beside each label in the legend.

```
> barplot(VADeaths[1:2,], angle=c(45, 135), density=30,
          col="black", names=c("RM", "RF", "UM", "UF"))
> legend(0.4, 38, c("55-59", "50-54"), cex=1.5,
         angle=c(135, 45), density=30)
```

It should be noted that it is entirely the responsibility of the user to ensure that the legend corresponds to the plot. There is no automatic checking that data symbols in the legend match those in the plot, or that the labels in the legend have any correspondence with the data. This is one area where the **lattice** and **ggplot2** graphics systems provide a significant convenience (see Part II).

Some high-level functions draw their own legend specific to their purpose (e.g., `filled.contour()`).

3.4.4   Axes

In most cases, the axes that are automatically generated by the base graphics system will be sufficient for a plot. This is true even when the data being plotted on an axis are not numeric. For example, the axes of a boxplot or barplot are labeled appropriately using group names (see Figure 3.20).

Section 3.2.5 describes ways in which the default appearance of automatically generated axes can be modified, but it is more often the case that the user needs to inhibit the production of the automatic axis and draw a customized axis using the `axis()` function.

The first step is to inhibit the default axes. Most high-level functions should provide an **axes** argument which, when set to FALSE, indicates that the high-

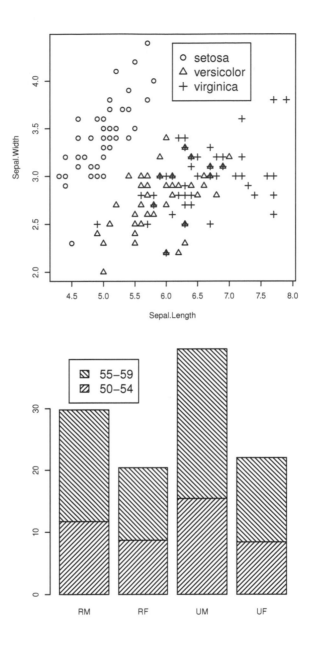

Figure 3.20
Some simple legends. Legends can be added to any kind of plot and can relate text labels to different symbols or different fill colors or patterns.

level function should not draw axes. Specifying the base graphics setting `xaxt="n"` (or `yaxt="n"`) may also do the trick.

The `axis()` function can draw axes on any side of a plot (chosen by the `side` argument), and the user can specify the location along the axis of tick marks and the text to use for tick labels (using the `at` and `labels` arguments, respectively). The following code demonstrates a simple example of a plot where the automatic axes are inhibited and custom axes are drawn, including a "secondary" y-axis on the right side of the plot (see Figure 3.21).*

First of all, a line plot is drawn with no axes.

```
> plot(nhtempCelsius, axes=FALSE, ann=FALSE, ylim=c(0, 13))
```

Next, the main y-axis is drawn with specific tick locations to represent the Centigrade scale. The number 2 means that the axis should be drawn in margin 2 (the left margin) and the `at` argument specifies the locations of the tick marks for the axis.

```
> axis(2, at=seq(0, 12, 4))
> mtext("Degrees Centigrade", side=2, line=3)
```

Now the default bottom axis is drawn and a secondary y-axis is drawn to represent the Fahrenheit scale. In the second expression, the `labels` argument is used to draw special tick mark labels on the secondary y-axis and this axis is drawn to the right of the plot by specifying 4 as the axis margin number.

```
> axis(1)
> axis(4, at=seq(0, 12, 4), labels=seq(0, 12, 4)*9/5 + 32)
> mtext(" Degrees Fahrenheit", side=4, line=3)
> box()
```

The `axis()` function is not generic, but there are special alternative functions for plotting time-related data. The functions `axis.Date()` and `axis.POSIXct()` take an object containing dates and produce an axis with appropriate labels representing times, days, months, and years (e.g., 10:15, Jan 12 or 1995).

In some cases, it may be useful to draw tick marks at the locations that the default axis would use, but with different labels. The `axTicks()` function can

---

*The data used in this plot are (a Celsius version of) mean annual temperature in degrees Fahrenheit in New Haven, Connecticut, from 1912 to 1971, available as the data set nhtemp in the **datasets** package.

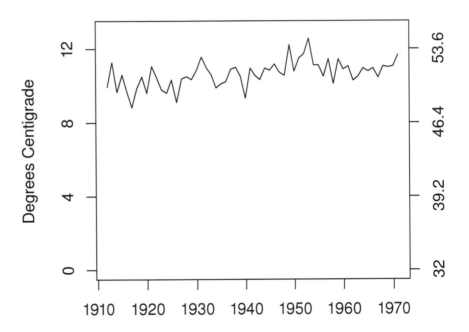

Figure 3.21

Customizing axes. On top is a data set drawn with the default axes. On the bottom, an initial plot is drawn with a y-scale in degrees Centigrade, including zero on the scale, then a secondary y-axis is drawn with a scale in degrees Fahrenheit. The labels on the secondary y-axis are specified explicitly, rather than just being the default numeric locations of the tick marks.

be used to calculate these default locations. This function is also useful for enforcing an `xaxp` (or `yaxp`) graphics state setting, which control the number and placement of tick marks. If these settings are specified via `par()`, they usually have no effect because the base graphics system almost always calculates the settings itself. The user can choose these settings by passing them as arguments to `axTicks()`, then passing the resulting locations via the `at` argument to `axis()`.

3.4.5   Coordinate systems

The base graphics system provides a number of coordinate systems for conveniently locating graphical output (see Section 3.1.1). Graphical output in the plot region is automatically positioned relative to the scales on the axes and text in the figure margins is placed in terms of a number of lines away from the edge of the plot (i.e., a scale that naturally corresponds to the size of the text).

It is also possible to locate output according to other coordinate systems that are not automatically supplied, but a little more work is required from the user. The basic principle is that the base graphics state can be queried to determine features of existing coordinate systems, then new coordinate systems can be calculated from this information.

The `par()` function

As well as being used to enforce new graphics state settings, the function `par()` can also be used to query current graphics state settings. The most useful settings are: `din`, `fin`, and `pin`, which reflect the current size, (`width`, `height`), of the graphics device, figure region, and plot region, in inches; and `usr`, which reflects the current user coordinate system (i.e., the ranges on the axes). The values of `usr` are in the order (`xmin`, `xmax`, `ymin`, `ymax`). When a scale has a logarithmic transformation, the values are (`10^xmin`, `10^xmax`, `10^ymin`, `10^ymax`).

There are also settings that reflect the size, (`width`, `height`), of a "standard" character. The setting `cin` gives the size in inches, `cra` in "rasters" or pixels, and `cxy` in "user coordinates." However, these values are not very useful because they only refer to a `cex` value of 1 (i.e., they ignore the current `cex` setting) *and* they only refer to the `ps` value when the current graphics device was first opened. Of more use are the `strheight()` function and the `strwidth()` function. These calculate the height and width of a given piece of text in inches, or in terms of user coordinates, or as a proportion of the current figure region (taking into account the current `cex` and `ps` settings).

Figure 3.22
Custom coordinate systems. The lines and text are drawn relative to real physical centimeters (rather than the default coordinate system defined by the scales on plot axes).

The following code demonstrates a simple example of making use of customized coordinates where a ruler is drawn showing centimeter units (see Figure 3.22).

A blank plot region is set up first and calculations are performed to establish the relationship between user coordinates in the plot and physical centimeters.*

```
> plot(0:1, 0:1, type="n", axes=FALSE, ann=FALSE)
> usr <- par("usr")
> pin <- par("pin")
> xcm <- diff(usr[1:2])/(pin[1]*2.54)
> ycm <- diff(usr[3:4])/(pin[2]*2.54)
```

Now drawing can occur with positions expressed in terms of centimeters. The ruler itself is drawn with a call to `rect()` to draw the edges of the ruler, a call to `segments()` to draw the scale, and calls to `text()` to label the scale.

```
> rect(0, 0, 1, 1, col="white")
> segments(seq(1, 8, 0.1)*xcm, 0,
            seq(1, 8, 0.1)*xcm,
          c(rep(c(0.5, rep(0.25, 4),
                   0.35, rep(0.25, 4)),
                7), 0.5)*ycm)
> text(1:8*xcm, 0.6*ycm, 0:7, adj=c(0.5, 0))
> text(8.2*xcm, 0.6*ycm, "cm", adj=c(0, 0))
```

---

*R graphics relies on having accurate information on the physical size of the natural units on the page or screen (e.g., the physical size of pixels on a computer screen). The physical size of output when producing PostScript and PDF files (see Section 9.1) should always be correct, but small inaccuracies may occur when specifying output with a physical size (such as inches) on a graphics window on screen.

Table 3.5

The coordinate systems recognized by the base graphics system.

Name	Description
"user"	The scales on the plot axes
"inches"	Inches, with (0, 0) at bottom-left
"device"	Pixels for screen or bitmap output, otherwise 1/72in.
"ndc"	Normalized coordinates, with (0, 0) at bottom-left and (1, 1) at top-right, within the entire device
"nic"	Normalized coordinates within the inner region
"nfc"	Normalized coordinates within the figure region
"npc"	Normalized coordinates within the plot region

There are utility functions, xinch() and yinch(), for performing the inches-to-user coordinates transformation (plus xyinch() for converting a location in one step and cm() for converting inches to centimeters). More powerful still are the grconvertX() and grconvertY() functions, which can be used to convert locations between any of the coordinate systems that the base graphics engine recognizes (see Table 3.5).

One problem with performing coordinate transformations like these is that the locations and sizes being drawn have no memory of how they were calculated. They are specified as locations and dimensions in user coordinates. This means that if the graphics window is resized (so that the relationship between physical dimensions and user coordinates changes), the locations and sizes will no longer have their intended meaning. If, in the above example, the graphics window is resized, the ruler will no longer accurately represent centimeter units. This problem will also occur if output is copied from one device to another device that has different physical dimensions. The legend() function performs calculations like these when arranging the components of a legend and its output is affected by resizing a device and copying between devices.*

Overlaying output

It is sometimes useful to plot two data sets on the same plot where the data sets share a common x-variable, but have very different y-scales. This can be achieved in at least two ways. One approach is simply to use par(new=TRUE) to overlay two distinct plots on top of each other, though care must be taken to avoid conflicting axes overwriting each other. Another approach is to explicitly

---

*It is possible to work around these problems in by using the recordGraphics() function, although this function should be used with extreme care.

reset the `usr` state setting before plotting a second set of data. The following code demonstrates both approaches to produce exactly the same result (see the top plot of Figure 3.23).

The data are yearly numbers of drunkenness-related arrests* and mean annual temperature in New Haven, Connecticut from 1912 to 1971. The temperature data are available as the data set `nhtemp` in the `datasets` package. There are only arrests data for the first 9 years.

```
> drunkenness <- ts(c(3875, 4846, 5128, 5773, 7327,
                      6688, 5582, 3473, 3186,
                      rep(NA, 51)),
                    start=1912, end=1971)
```

The first approach is to draw a plot of the drunkenness data, call `par(new=TRUE)`, then draw a complete second plot of the temperature data on top of the first plot. The second plot does not draw default axes (`axes=FALSE`), but uses the `axis()` function to draw a secondary y-axis to represent the temperature scale.

```
> par(mar=c(5, 6, 2, 4))
> plot(drunkenness, lwd=3, col="gray", ann=FALSE, las=2)
> mtext("Drunkenness\nRelated Arrests", side=2, line=3.5)
> par(new=TRUE)
> plot(nhtemp, ann=FALSE, axes=FALSE)
> mtext("Temperature (F)", side=4, line=3)
> title("Using par(new=TRUE)")
> axis(4)
```

The second approach draws only one plot (for the drunkenness data). The user coordinate system is then redefined by specifying a new `usr` setting and the second "plot" is produced simply using `lines()`. Again, a secondary axis is drawn using the `axis()` function.

---

*These data were obtained as "Crime Statistics and Department Demographics" from the New Haven Police Department:
http://www.cityofnewhaven.com/police/html/stats/crime/yearly/1863-1920.htm.

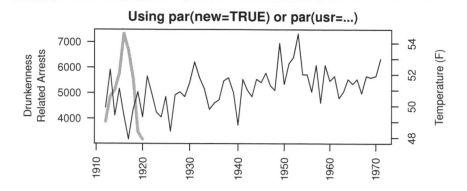

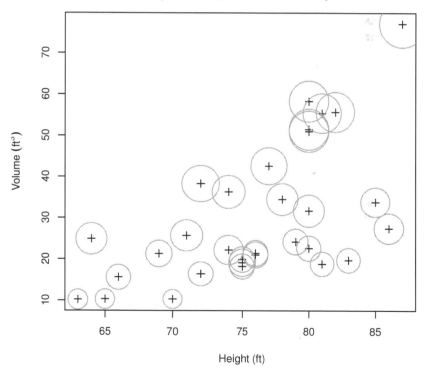

Figure 3.23
Overlaying plots. In the top plot, two line plots are drawn one on top of the other
to produce aligned plots of two data sets with very different scales. In the bottom
plot, the plotting function `symbols()` is used in "annotating mode" so that it adds
circles to an existing scatterplot rather than producing a complete plot itself.

```
> par(mar=c(5, 6, 2, 4))
> plot(drunkenness, lwd=3, col="gray", ann=FALSE, las=2)
> mtext("Drunkenness\nRelated Arrests", side=2, line=3.5)
> usr <- par("usr")
> par(usr=c(usr[1:2], 47.6, 54.9))
> lines(nhtemp)
> mtext("Temperature (F)", side=4, line=3)
> title("Using par(usr=...)")
> axis(4)
```

Some high-level functions (e.g., `symbols()` and `contour()`) provide an argument called **add** which, if set to **TRUE**, will add the function output to the current plot, rather than starting a new plot. The following code shows the `symbols()` function being used to annotate a basic scatterplot (see the bottom plot of Figure 3.23). The data used in this example are physical measurements of black cherry trees available as the **trees** data frame from the **datasets** package.

```
> with(trees,
      {
        plot(Height, Volume, pch=3,
             xlab="Height (ft)",
             ylab=expression(paste("Volume ", (ft^3))))
        symbols(Height, Volume, circles=Girth/12,
                fg="gray", inches=FALSE, add=TRUE)
      })
```

Another function of this type is the `bxp()` function. This function is called by `boxplot()` to draw the individual boxplots and is specifically set up to add boxplots to an existing plot (although it can also produce a complete plot).

It is also worth remembering that R follows a painters model, with later output obscuring earlier output. The following example makes use of this feature to fill a complex region within a plot (see Figure 3.24).

The first step is to prepare the data and calculate some important features of the data.

```
> x <- as.numeric(time(nhtemp))
> y <- as.numeric(nhtemp)
> n <- length(x)
> mean <- mean(y)
```

The first thing to draw is a plot with a filled polygon beneath the y-values (see the top-left plot of Figure 3.24).

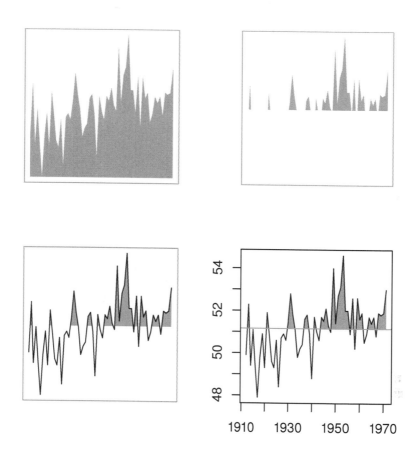

Figure 3.24
Overlaying output (making use of the painters model). The final complex plot, shown at bottom-right, is the result of overlaying several basic pieces of output: a gray polygon at top-left, with a white rectangle over the top (top-right), a black line on top of that (bottom-left), and a gray line on top of it all (plus axes and a bounding box).

```
> plot(x, y, type="n", axes=FALSE, ann=FALSE)
> polygon(c(x[1], x, x[n]), c(min(y), y, min(y)),
          col="gray", border=NA)
```

The next step is to draw a rectangle over the top of the polygon up to a fixed y-value. The expression `par("usr")` is used to obtain the current x-scale and y-scale ranges (see the top-right plot of Figure 3.24).

```
> usr <- par("usr")
> rect(usr[1], usr[3], usr[2], mean, col="white", border=NA)
```

Now a line through the y-values is drawn over the top of the rectangle (see the bottom-left plot of Figure 3.24).

```
> lines(x, y)
```

Finally, a horizontal line is drawn to indicate the y-value cut-off, and axes are added to the plot (see the bottom-right plot of Figure 3.24).

```
> abline (h=mean, col="gray")
> box()
> axis(1)
> axis(2)
```

3.4.6   Special cases

Some high-level functions are a little more difficult to annotate than others because the plotting regions that they set up either are not immediately obvious or are not available after the function has run. This section describes a number of high-level functions where additional knowledge is required to perform annotations.

Obscure scales on axes

It is not immediately obvious how to add extra annotation to a barplot or a boxplot in base R graphics because the scale on the categorical axis is not obvious.

The difficulty with the `barplot()` function is that, because the scale on the x-axis is not labeled at all by default, the numeric scale is not obvious (and calling `par("usr")` is not much help because the scale that the function sets up is not intuitive either). In order to add annotations sensibly to a barplot it is necessary to capture the value returned by the function. This return value gives the x-locations of the mid-points of each bar that the function has drawn. These midpoints can then be used to locate annotations relative to the bars in the plot.

The code below shows an example of adding extra horizontal reference lines to the bars of a barplot. The mid-points of the bars are saved to a variable called `midpts`, then locations are calculated from those mid-points (and the original counts) to draw horizontal white line segments within each bar using the `segments()` function (see the left plot of Figure 3.25).

```
> y <- sample(1:10)
> midpts <- barplot(y, col=" light gray")
> width <- diff(midpts[1:2])/4
> left <- rep(midpts, y - 1) - width
> right <- rep(midpts, y - 1) + width
> heights <- unlist(apply(matrix(y, ncol=10),
                   2, seq))[-cumsum(y)]
> segments(left, heights, right, heights,
         col="white")
```

The `boxplot()` function is similar to the `barplot()` function in that the x-scale is typically labeled with category names so the numeric scale is not obvious from looking at the plot. Fortunately, the scale set up by the `boxplot()` function is much more intuitive. The individual boxplots are drawn at x-locations `1:n`, where `n` is the number of boxplots being drawn.

The following code provides a simple example of annotating boxplots to add a jittered dotplot of individual data points on top of the boxplots. This provides a detailed view of the data with individual points and shows the main features of the data via the boxplot. It is also a useful way to show how interesting features of the data, such as small clusters of points, can be hidden by a boxplot. In this example, the jittered data are centered upon the x-locations `1:2` to correspond to the centers of the relevant boxplots (see the right plot of Figure 3.25).

```
> with(ToothGrowth,
     {
       boxplot(len ~ supp, border="gray",
               col="light gray", boxwex=0.5)
       points(jitter(rep(1:2, each=30), 0.5),
              unlist(split(len, supp)),
              cex=0.5, pch=16)
     })
```

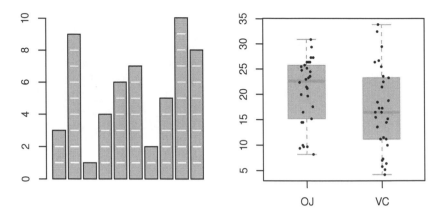

Figure 3.25
Special-case annotations. Some examples of functions where annotation requires
special care. In the barplot at left, the value returned by the `barplot()` function is
used to add horizontal white lines within the bars. Jittered points are added to the
boxplot (right) using the knowledge that the *i*th box is located at position *i* on the
x-axis.

Functions that draw several plots

The `pairs()` function is an example of a high-level function that draws more
than one plot. This function draws a matrix of scatterplots. Such functions tend to save the base graphics state before drawing, call `par(mfrow)` or
`layout()` to arrange the individual plots, and restore the base graphics state
once all of the individual plots have been drawn. This means that it is not
possible to annotate any of the plots drawn by the `pairs()` function once the
function has completed drawing. The regions and coordinate systems that the
function set up to draw the individual plots have been thrown away. The only
way to annotate the output from such functions is by way of *panel functions*.

The `pairs()` function has a number of arguments that allow the user to
specify a function: `panel`, `diag.panel`, `upper.panel`, `lower.panel`, and
`text.panel`. The functions specified via these arguments are run as each
individual plot is drawn. In this way, the panel function has access to the plot
regions that are set up for each individual plot.

The following code shows a `pairs()` plot of the first two variables in the `iris`
data set. The `diag.panel` argument is used to draw boxplots in the diagonal
panels, instead of the default variable names. Notice that the panel function
must only add extra output, *not* start its own plot and this is achieved in this

case by called `boxplot()` with `add=TRUE`. Because `axes=FALSE`, the normal boxplot axes are not drawn, and the `at` argument is used to make sure the boxplots are centered horizontally within the panels. Because the normal diagonal panels have variable names drawn in them, a `text.panel` function is also specified. This panel function calls `mtext()` so that the normal text is drawn in the top margin of the panel instead. The resulting plot is shown in Figure 3.26.

```
> pairs(iris[1:2],
        diag.panel=function(x, ...) {
            boxplot(x, add=TRUE, axes=FALSE,
                    at=mean(par("usr")[1:2]))
        },
        text.panel=function(x, y, labels, ...) {
            mtext(labels, side=3, line=0)
        })
```

The `filled.contour()` function and the `coplot()` function have the same problem as `pairs()` because the legends that they draw are actually separate plots. Again, those functions allow annotation via panel function arguments.

The `panel.smooth()` function provides a predefined panel function to add a smoothed trend line to a scatterplot of points.

3D plots

It is possible to annotate a plot that was produced using the `persp()` function, but it is more difficult than for most other high-level functions. The important step is to acquire the transformation matrix that the `persp()` function returns. This can be used to transform 3D locations into 2D locations, using the `trans3d()` function. The result can then be given to the standard annotation functions such as `lines()` and `text()`. The `persp()` function also has an `add` argument, which allows multiple `persp()` plots to be over-plotted.

The following code demonstrates annotation of `persp()` output to add a contour plot beneath a 3D plot of the Maunga Whau volcano in Auckland New Zealand (see Figure 3.27). The data are from the `volcano` matrix in the **datasets** package.

The first step is to draw the 3D surface. The important features of this code are that the `zlim` is specified to leave room for the contour plot and the result of the call to `persp()` is assigned to a variable called **trans**.

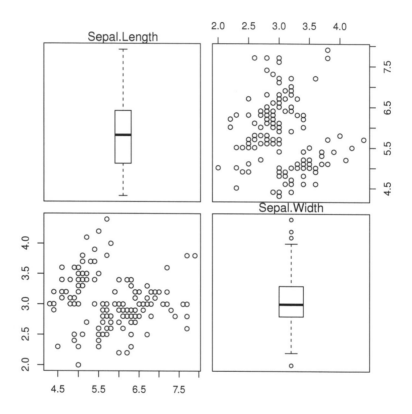

Figure 3.26

A panel function example. An example of using a panel function to add customized output to each of the diagonal panels of a `pairs()` plot.

```
> z <- 2 * volcano
> x <- 10 * (1:nrow(z))
> y <- 10 * (1:ncol(z))
> trans <- persp(x, y, z, zlim=c(0, max(z)),
                 theta = 150, phi = 12, lwd=.5,
                 scale = FALSE, axes=FALSE)
```

The next code calculates contour lines from the 3D data and then adds them to the plot. The result of `contourLines()` is a list, so `lapply()` is used to draw each contour line separately. The locations of the contour lines in the 3D plot are calculated using `trans3d()`, which is given the x and y vertices for a contour line, plus the z-position of zero (below the 3D surface). The `trans3d()` function converts the 3D locations into 2D locations which are drawn with the `lines()` function.

```
> clines <- contourLines(x, y, z)
> lapply(clines,
         function(contour) {
             lines(trans3d(contour$x, contour$y, 0, trans))
         })
```

A major limitation with annotating `persp()` output is that there is no support for automatically hiding output that should not be seen. In the above example, the view point was carefully chosen so that the entire contour plot was visible beneath the 3D surface. If the viewing angle is changed so that the surface and the contour lines overlap, the contour lines will be drawn *on top of* the 3D surface because they are drawn second. In simple cases, this sort of problem can be worked around through careful ordering of drawing operations, but in the general case a more sophisticated 3D graphics system would be required (e.g., the **rgl** package).

3.5   Creating new plots

There are cases where no existing plot provides a sensible starting point for creating the final plot that the user requires; situations where simply drawing more shapes on the plot is not sufficient. This section describes how to construct a new plot entirely from scratch for such cases.

The `plot.new()` function is the most basic starting point for producing a base graphics plot (the `frame()` function is equivalent). This function starts a new

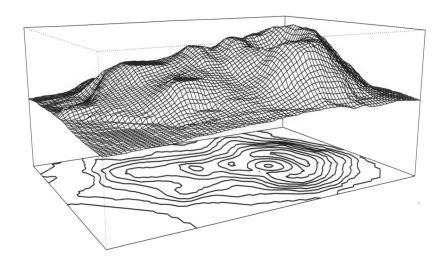

Figure 3.27
Annotating a 3D surface created by `persp()`. The contour lines are added to the
3D plot using the transformation matrix returned by the `persp()` function.

plot and sets up the various plotting regions described in Section 3.1.1, with
both the x-scale and y-scale set to $(0, 1)$.* The size and position of the regions
that are set up depend on the current graphics state settings (see Section
3.2.6).

The `plot.window()` function resets the scales in the user coordinate system,
given x- and y-ranges via the arguments `xlim` and `ylim`, and the `plot.xy()`
function draws data symbols and lines between locations within the plot region.

3.5.1   A simple plot from scratch

In order to demonstrate the use of these functions, the following code produces
a very simple scatterplot like Figure 1.1 from scratch. The result is shown in
Figure 3.28.

---

*The actual scale setup depends on the current settings for `xaxs` and `yaxs`. With the
default settings, the scales are $(-0.04, 1.04)$.

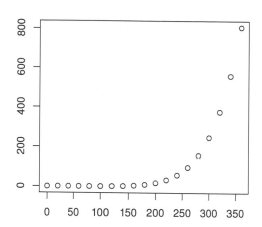

Figure 3.28

A simple scatterplot of vapor pressure of mercury as a function of temperature. This is similar to Figure 1.1, but where that figure was generated with a single call to the **plot()** function, this plot is produced from scratch using low-level plotting functions.

```
> plot.new()
> plot.window(range(pressure$temperature),
               range(pressure$pressure))
> plot.xy(pressure, type="p")
> box()
> axis(1)
> axis(2)
```

The call to **plot.new()** starts a new, completely blank, plot and the call to **plot.window()** sets the scales on the axes to fit the range of the data to be plotted. At this point, there is still nothing drawn. The **plot.xy()** function draws data symbols (**type="p"**) at the data locations, then **box()** draws a rectangle around the plot region, and **axis()** is used to draw the axes.

The output could be produced by the simple expression **plot(pressure)**, but this code shows that the steps in building a plot are available as separate functions as well, which allows the user to have fine control over the construction of a plot.

3.5.2   A more complex plot from scratch

This section describes a slightly more complex example of creating a plot from scratch. The final goal is represented in Figure 3.29 and the steps involved are described below.

The first chunk of code prepares some data to plot. These are the counts of (adult) male and female survivors of the sinking of the *Titanic*.

```
> groups <- dimnames(Titanic)[[1]]
> males <- Titanic[, 1, 2, 2]
> females <- Titanic[, 2, 2, 2]

> males

 1st  2nd  3rd Crew
  57   14   75  192

> females

 1st  2nd  3rd Crew
 140   80   76   20
```

There are several ways that the plot could be created, the main idea being that it fundamentally consists of just a collection of graphical primitives that have been arranged in a meaningful way.

For this example, the approach will be to create a single plot. The labels to the left of the plot will be drawn in the margins of the plot, but everything else will be drawn inside the plot region. This next bit of code sets up the figure margins so that there is enough room for the labels in the left margin, but all other margins are nice and small (to avoid lots of empty space around the plot).

```
> par(mar=c(0.5, 3, 0.5, 1))
```

Inside the plot region there are six different rows of output to draw: the four main pairs of bars, the x-axis, and the legend at the bottom. The axis will be drawn at a y-location of 0, the main bars at the y-locations 1:4, and the legend at -1. The following code starts the plot and sets up the appropriate y-scale and x-scale.

```
> plot.new()
> plot.window(xlim=c(-200, 200), ylim=c(-1.5, 4.5))
```

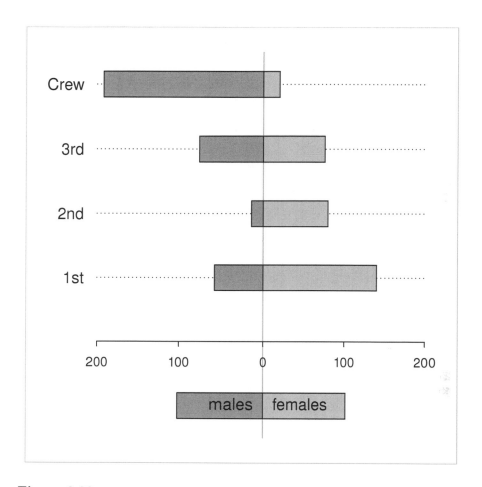

Figure 3.29
A back-to-back barplot from scratch. This demonstrates the use of lower-level plotting functions to produce a novel plot that cannot be produced by an existing high-level function.

This next bit of code assigns some useful values to variables, including the x-locations of tick marks on the x-axis, the y-locations of the main bars, and a value representing half the height of the bars.

```
> ticks <- seq(-200, 200, 100)
> y <- 1:4
> h <- 0.2
```

Now some drawing can occur. This next code draws the main part of the plot. Everything is drawn using calls to the low-level functions such as `lines()`, `segments()`, `mtext()`, and `axis()`. In particular, the main bars are just rectangles produced using `rect()`. Notice that the x-axis is drawn within the plot region (pos=0).

```
> lines(rep(0, 2), c(-1.5, 4.5), col="gray")
> segments(-200, y, 200, y, lty="dotted")
> rect(-males, y-h, 0, y+h, col="dark gray")
> rect(0, y-h, females, y+h, col="light gray")
> mtext(groups, at=y, adj=1, side=2, las=2)
> par(cex.axis=0.8, mex=0.5)
> axis(1, at=ticks, labels=abs(ticks), pos=0)
```

The final step is to produce the legend at the bottom of the plot. Again, this is just a series of calls to low-level functions, although the bars are sized using `strwidth()` to ensure that they contain the labels.

```
> tw <- 1.5*strwidth("females")
> rect(-tw, -1-h, 0, -1+h, col="dark gray")
> rect(0, -1-h, tw, -1+h, col="light gray")
> text(0, -1, "males", pos=2)
> text(0, -1, "females", pos=4)
```

This example is particularly customized to the data set involved. It could be made much more general by replacing some constants with variable values (e.g., instead of using 4 because there are four groups in the data set, the code could have a variable `numGroups`). If more than one such plot needs to be made, it makes good sense to also wrap the code within a function. That task is discussed in the next section.

3.5.3   Writing base graphics functions

Having made the effort to construct a plot from scratch, the next step is to encapsulate the calls within a new function and possibly even make it available for others to use. This section briefly describes some of the things to consider when creating a new graphics function built on the base graphics system.

There are many advantages to developing new graphics functions in the **grid** graphics system (see Part II) rather than using base graphics. See Chapter 8 for a more complete discussion of the issues involved in developing new graphics functions.

Helper functions

There are some helper functions that do no drawing, but are used by the predefined high-level plots to do some of the work in setting up a plot.

The `xy.coords()` function is useful for allowing `x` and `y` arguments to your new function to be flexibly specified (just like the `plot()` function where `y` can be left unspecified and `x` can be a `data.frame`, and so on). This function takes `x` and `y` arguments and creates a standard object containing x-values, y-values, and sensible labels for the axes. There is also an `xyz.coords()` function for plots of three variables.

If your plotting function generates multiple subplots, the `n2mfrow()` function may be helpful to generate a sensible number of rows and columns of plots, based on the total number of plots to fit on a page.

Another set of useful helper functions are those that calculate values to plot from the raw data (but do no actual drawing). Examples of these sorts of functions are: `boxplot.stats()` used by `boxplot()` to generate five-number summaries; `contourLines()` used by `contour()` to generate contour lines; `nclass.Sturges()`, `nclass.scott()`, and `nclass.FD()` used by `hist()` to generate the number of intervals for a histogram; and `co.intervals()` used by `coplot()` to generate ranges of values for conditioning a data set into panels.

Some high-level functions invisibly return this sort of information too. For example, `boxplot()` returns the combined results from `boxplot.stats()` for all of the boxplots that it produces, and `hist()` returns information on the intervals that it creates including the number of data values in each interval. The `hist()` function is also useful (with `plot=FALSE`) simply to perform binning of continuous data.

Argument lists

A common technique when writing a base graphics function is to provide an ellipsis argument (. . .) instead of individual graphics state arguments (such as `col` and `lty`). This allows users to specify any state settings (e.g., `col="red"` and `lty="dashed"`) and the new function can pass them straight on to the base graphics functions that the new function calls. This avoids having to specify all individual state settings as arguments to the new function. Some care must be taken with this technique because sometimes different graphics functions interpret the same graphics state setting in different ways (the `col` setting is a good example; see Section 3.2). In such cases, it becomes necessary to name the individual graphics state setting as an argument and explicitly pass it on only to other graphics calls that will accept it and respond to it in the desired manner.

Sometimes it is useful for a graphics function to deliberately override the current graphics state settings. For example, a new plot may want to force the `xpd` setting to be `NA` in order to draw lines and text outside of the plot region. In such cases, it is polite for the graphics function to revert the graphics state settings at the end of the function so that users do not get a nasty surprise! A standard technique is to put the following expressions at the start of the new function to restore the graphics state to the settings that existed before the function was called.

```
opar <- par(no.readonly=TRUE)
on.exit(par(opar))
```

Because some of the base graphics state settings interact with each other, such a wholesale save-and-replace approach is actually unlikely to return the graphics state to exactly what it was before, so an even better solution is to save and restore only those parameters that the function modifies.

Care should be taken to ensure that a new graphics function takes notice of appropriate graphics state settings (e.g., `ann`). This can be a little complicated to implement because it is necessary to be aware of the possibility that the user might specify a setting in the call to the function and that such a setting should override the main graphics state setting. The standard approach is to name the state setting explicitly as an argument to the graphics function and provide the permanent state setting as a default value. See the new graphics function template below for an example of this technique using the `ann` argument. An additional complication is that now there is a state setting that will not be part of the ... argument, so the state setting must be explicitly passed on to any other functions that might make use of it.

Another good technique is to provide arguments that users are used to seeing in other graphics functions — the `main`, `sub`, `xlim`, and `ylim` arguments are good examples of this sort of thing — and a new graphics function should be able to handle missing and non-finite values. The functions `is.na()`, `is.finite()`, and `na.omit()` may be useful for this purpose.

Plot methods

If a new function is for use with a particular type of data, then it is convenient for users if the function is provided as a method for the generic `plot()` function. This allows users to simply call the new function by calling `plot(x)`, where x is an object of the relevant class.

A graphics function template

The code in Figure 3.30 is a simple shell that combines some of the basic guidelines from this section. This is just a simplified version of the default `plot()` method. It is far from complete and will not gracefully accept all possible inputs (especially via the ... argument), but it could be used as the starting template for writing a new base graphics function.

3.6   Interactive graphics

The strength of the base graphics system lies in the production of static graphics and that is the focus of this book. However, for completeness, this section briefly mentions the limited facilities for interacting with base graphics output.

The `locator()` function allows the user to click within a plot and returns the coordinates where the mouse click occurred. It will also optionally draw data symbols at the clicked locations or draw lines between the clicked locations.

The `identify()` function can be used to add labels to data symbols on a plot. The data point closest to the mouse click gets labeled.

There is also a more general-purpose mechanism for defining interactions with the output in a graphics window (though at the time of writing only for the Windows, X Window, and Cairo graphics devices; see Chapter 9). The `setGraphicsEventHandlers()` function can be used to define R functions that will be called whenever events such as keystrokes or mouse clicks occur within the graphics window and the `getGraphicsEvent()` function can be called to start listening to events within the graphics window. This provides a more flexible basis for developing simple interactive base graphicsplots.

```
 1 plot.newclass <- function(x, y=NULL,
 2                            main="", sub="",
 3                            xlim=NULL, ylim=NULL,
 4                            axes=TRUE, ann=par("ann"),
 5                            col=par("col"),
 6                            ...) {
 7     xy <- xy.coords(x, y)
 8     if (is.null(xlim))
 9         xlim <- range(xy$x[is.finite(xy$x)])
10     if (is.null(ylim))
11         ylim <- range(xy$y[is.finite(xy$y)])
12     opar <- par(no.readonly=TRUE)
13     on.exit(par(opar))
14     plot.new()
15     plot.window(xlim, ylim, ...)
16     points(xy$x, xy$y, col=col, ...)
17     if (axes) {
18         axis(1)
19         axis(2)
20         box()
21     }
22     if (ann)
23         title(main=main, sub=sub,
24               xlab=xy$xlab, ylab=xy$ylab, ...)
25 }
```

Figure 3.30
A graphics function template. This code provides a starting point for producing a
new graphics function for others to use.

Chapter summary

High-level base graphics functions produce complete plots, and low-level base graphics functions add output to existing plots. There are low-level functions for producing simple output such as lines, rectangles, text, and polygons and also functions for producing more complex output such as axes and legends.

The base graphics system creates several regions for drawing the various components of a plot: a plot region for drawing data symbols and lines, figure margins for axes and labels, and so on. Each low-level graphics function produces output in a particular drawing region and most work in the plot region.

There is a base graphics system state that consists of settings to control the appearance of output and the arrangement of the drawing regions. There are settings for controlling color, fonts, line styles, data symbol style, and the style of axes. There are several mechanisms for arranging multiple plots on a single page.

It is straightforward to create a complete plot using only low-level graphics functions. This makes it possible to produce a completely new type of plot. It is also possible for the user to define an entirely new graphics function.

Part II

GRID GRAPHICS

4

Trellis Graphics: The lattice Package

Chapter preview

This chapter describes how to produce plots using the **lattice** package. There is a description of what **lattice** plots are as well as a description of the functions used to produce them. Plots produced by the **lattice** package are designed to be clear and easy to interpret and at the same time provide sophisticated plotting styles, such as multipanel conditioning. The **grid** graphics system provides no high-level plotting functions itself, so this chapter also describes one way to produce a complete plot using the **grid** system.

This part of the book concerns the major graphics packages that are related to the **grid** graphics system. This graphics system exists in parallel with the base graphics system and the two worlds do not naturally interact well (see Section 1.2, but also Chapter 12).

The **grid** package only provides low-level graphics functions; it does not provide any functions for drawing complete plots. Such high-level functions are provided instead by other packages. This chapter and the next describe two major packages of this type: Deepayan Sarkar's **lattice** and Hadley Wickham's **ggplot2**.

The **lattice** package implements Bill Cleveland's Trellis Graphics system with some novel extensions. This represents a complete and coherent graphics system, which can in most cases be used without encountering any concepts of the underlying **grid** system.

This chapter deals with **lattice** as a self-contained system consisting of functions for producing complete plots and functions for controlling the appearance of the plots. Section 6.8 and Section 7.14 describe some of the benefits that can be gained from viewing **lattice** plots as **grid** output and dealing directly with the **grid** concepts and objects that underlie the **lattice** system.

The graphics functions that make up the **lattice** graphics system are provided in an extension package called **lattice**, which is loaded into R as follows.

```
> library(lattice)
```

This chapter provides a very brief introduction to **lattice**. Much more information can be obtained from Deepayan Sarkar's book, "Lattice: Multivariate Data Visualization with R."

4.1   The lattice graphics model

In simple usage, **lattice** functions appear to work just like base graphics functions where the user calls a function and output is generated on the current device. The following code produces the **lattice** equivalent of the base graphics call `plot(pressure)`. The first argument is a formula defining the x- and y-variables to plot and the second argument is a data frame that contains the variables named in the formula. The resulting plot (Figure 4.1) should be compared with Figure 1.1.

```
> xyplot(pressure ~ temperature, pressure)
```

There are also many familiar arguments to modify the basic features of a **lattice** plot. For example, the following code plots lines as well as points, using the `type` argument, adds a title, using the `main` argument, and uses `pch` and `lty` to set the data symbol and line type (see Figure 4.2).

```
> xyplot(pressure ~ temperature, pressure,
         type="o", pch=16, lty="dashed",
         main="Vapor Pressure of Mercury")
```

Adding further lines and text to a plot is a little more complex in **lattice** compared to base graphics, so that topic is discussed later in Section 4.7.

Another important difference compared to base graphics functions is that **lattice** graphics functions do not produce graphical output directly. Instead

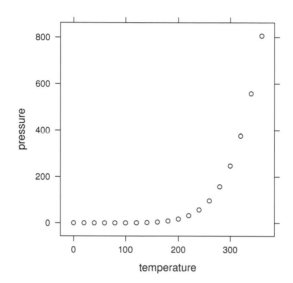

Figure 4.1

A scatterplot using **lattice** (showing the vapor pressure of mercury as a function of temperature). A basic **lattice** plot has a very similar appearance to an analogous base plot.

Vapor Pressure of Mercury

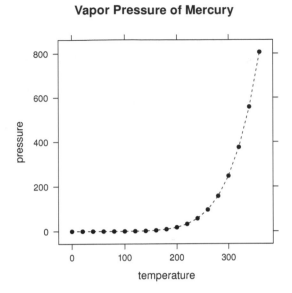

Figure 4.2
A modified scatterplot using **lattice**. Many of the standard high-level base graphics
arguments also work with **lattice**.

they produce an object of class **"trellis"**, which contains a description of the
plot. The **print()** method for objects of this class does the actual drawing of
the plot. This can be demonstrated quite easily. For example, the following
code creates a **trellis** object, but does not draw anything.

```
> tplot <- xyplot(pressure ~ temperature, pressure)
```

The result of the call to **xyplot()** is assigned to the variable **tplot** so it is
not printed. The plot can be drawn by calling print on the **trellis** object
(the result is exactly the same as Figure 4.1).

```
> print(tplot)
```

This explicit printing is necessary when calling **lattice** functions within a loop
or from another function.

4.1.1   Why another graphics system?

A number of functions in **lattice** produce output that is very similar to the output of functions in the base graphics system, but there are several reasons for using **lattice** functions instead of the base counterparts:

- The default appearance of the **lattice** plots is superior in some areas. For example, the default colors and the default data symbols have been deliberately chosen to make it easy to distinguish between groups when more than one data series is plotted, based on visual perception experiments. There are also some subtle things such as the fact that tick labels on the y-axes are written horizontally by default, which makes them easier to read.

- The arrangement of plot components is more automated in **lattice**. For example, the right amount of space is automatically created for axis labels and the plot title (it is usually not necessary to set figure margins manually).

- Legends can be automatically generated by the **lattice** system, so it is not the user's responsibility to ensure that the content of the legend corresponds correctly to the colors and data symbols used in the plot.

- The **lattice** plot functions can be extended in several very powerful ways. For example, several data series can be plotted at once in a convenient manner and multiple panels of plots can be produced easily (see Section 4.3).

- The output from **lattice** functions is **grid** output, so many powerful **grid** features are available for annotating, editing, and saving the graphics output. See Sections 6.8 and 7.14 for examples of these features.

4.2   lattice plot types

The **lattice** package provides functions to produce a number of standard plot types, plus some more modern and specialized plots. Table 4.1 describes the functions that are available and Figure 4.3 provides a basic idea of the sort of output that they produce.

Most of the **lattice** plotting functions provide a very long list of arguments and produce a wide range of different types of output. However, because **lattice** provides a single coherent system, many of the arguments are the

Table 4.1
The plotting functions available in **lattice**.

lattice Function	Description	Base Analog
barchart()	Barcharts	barplot()
bwplot()	Boxplots Box-and-whisker plots	boxplot()
densityplot()	Conditional kernel density plots Smoothed density estimate	plot.density
dotplot()	Dotplots Continuous vs. categorical	dotchart()
histogram()	Histograms	hist()
qqmath()	Quantile–quantile plots Data set vs. theoretical distribution	qqnorm()
stripplot()	Stripplots One-dimensional scatterplot	stripchart()
qq()	Quantile–quantile plots Data set vs. data set	qqplot()
xyplot()	Scatterplots	plot()
levelplot()	Level plots	image()
contourplot()	Contour plots	contour()
cloud()	3D scatterplot	-
wireframe()	3D surfaces	persp()
splom()	Scatterplot matrices	pairs()
parallelplot()	Parallel coordinate plots	-

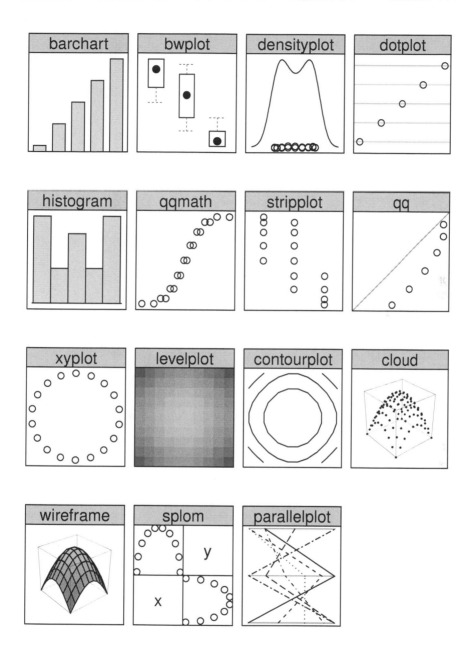

Figure 4.3
Plot types available in **lattice**. The name of the function used to produce the
different plot types is shown in the strip above each plot.

same across the different graphics functions, so much can be learned from just studying one of the **lattice** functions. This chapter will largely focus on the `xyplot()` function.

The following sections address the most important shared arguments. For a full explanation of all arguments, the help documentation should be consulted, particularly the help for the `xyplot()` function. The purpose of this chapter is to provide enough information to produce a range of complete plots using **lattice**.

4.3   The `formula` argument and multipanel conditioning

In most cases, the first argument to the **lattice** plotting functions is an R formula that describes which variables to plot. The simplest case has already been demonstrated. A formula of the form `y ~ x` plots variable `y` against variable `x`. There are some variations for plots of only one variable or plots of more than two variables. For example, for the `histogram()` function, the formula can be of the form `~ x` and for the `cloud()` and `wireframe()` functions something of the form `z ~ x * y` is required to specify the three variables to plot. Another useful variation is the ability to specify multiple y-variables. Something of the form `y1 + y2 ~ x` produces a plot of both the `y1` variable and the `y2` variable against `x`. Multiple x-variables can be specified as well.

The second argument to a **lattice** plotting function is typically `data`, which allows the user to specify a data frame within which **lattice** can find the variables that were used in the formula.

One of the very powerful features of Trellis Graphics is the ability to specify conditioning variables within the formula argument. Something of the form `y ~ x | g` indicates that several plots should be generated, showing the variable `y` against the variable `x` for each level of the variable `g`.

The following examples use various measurements on 32 different automobile designs, which are available as the data set `mtcars` in the `datasets` package. The examples will use measurements on fuel efficiency in miles per gallon (`mpg`), engine size or displacement (`disp`), and number of forward gears (`gear`).

```
> head(mtcars)
```

	mpg	cyl	disp	hp	drat	wt	qsec	vs	am
Mazda RX4	21.0	6	160	110	3.90	2.620	16.46	0	1
Mazda RX4 Wag	21.0	6	160	110	3.90	2.875	17.02	0	1
Datsun 710	22.8	4	108	93	3.85	2.320	18.61	1	1
Hornet 4 Drive	21.4	6	258	110	3.08	3.215	19.44	1	0
Hornet Sportabout	18.7	8	360	175	3.15	3.440	17.02	0	0
Valiant	18.1	6	225	105	2.76	3.460	20.22	1	0

	gear	carb
Mazda RX4	4	4
Mazda RX4 Wag	4	4
Datsun 710	4	1
Hornet 4 Drive	3	1
Hornet Sportabout	3	2
Valiant	3	1

A simple scatterplot of fuel efficiency as a function of engine size is produced by the following code (see Figure 4.4).

```
> xyplot(mpg ~ disp, data=mtcars)
```

As an example of multipanel conditioning, the following code produces several scatterplots, with each scatterplot showing the relationship between engine size and fuel efficiency for cars with a particular number of forward gears (see Figure 4.5).

```
> xyplot(mpg ~ disp | factor(gear), data=mtcars)
```

In the Trellis terminology, the plot in Figure 4.5 consists of three *panels*. Each panel in this case contains a scatterplot and above each panel there is a *strip* that presents the level of the conditioning variable. There can be more than one conditioning variable in the formula argument, in which case a panel is produced for each combination of the conditioning variables.

The most natural type of variable to use as a conditioning variable is a categorical variable (factor), but there is also support for using a continuous (numeric) conditioning variable. For this purpose, Trellis Graphics introduces the concept of a *shingle*. This is a continuous variable with a number of ranges associated with it. The ranges are used to split the continuous values into (possibly overlapping) groups. The **shingle()** function can be used to explicitly control the ranges, or the **equal.count()** function can be used to generate ranges automatically given a number of groups.

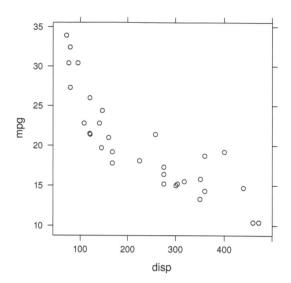

Figure 4.4

A **lattice** scatterplot of fuel efficiency as a function of engine size.

4.4 The group argument and legends

Another important argument in high-level **lattice** functions is the **group** argument, which allows multiple data series to be drawn on the same plot (or in each panel). The following code shows an example and the result is shown in Figure 4.6.

```
> xyplot(mpg ~ disp, data=mtcars,
         group=gear,
         auto.key=list(space="right"))
```

By specifying a variable via the **group** argument, a different plotting symbol will be used for cars with different numbers of gears. The **auto.key** argument is set so that **lattice** automatically generates an appropriate legend to show the mapping between data symbols and number of gears. This argument can either be just **TRUE** or a list of values specifying the appearance of the legend. In this case, the legend is positioned to the right of the plot. Notice that the page is automatically arranged to provide space for the plot legend.

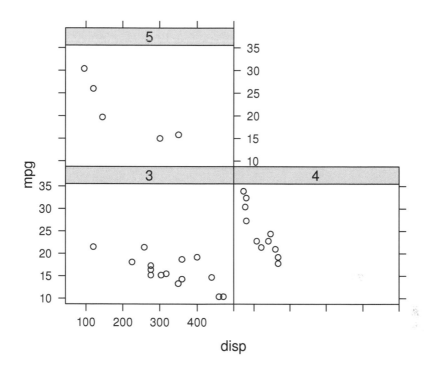

Figure 4.5
A **lattice** multipanel conditioning plot. A single function call produces several scatterplots of the relationship between engine size and fuel efficiency for cars with different numbers of forward gears.

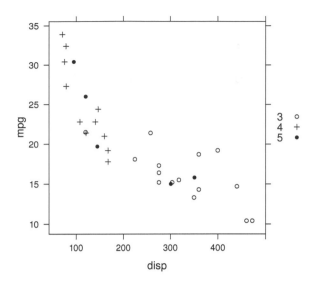

Figure 4.6
A **lattice** plot with multiple groups and an automatically generated legend. Different data symbols are used for cars with different numbers of gears.

In addition to the `auto.key`, there are arguments `key` and `legend` which provide progressively greater flexibility at the cost of increased complexity.

4.5 The `layout` argument and arranging plots

There are two types of arrangements to consider when dealing with **lattice** plots: the arrangement of panels and strips within a single **lattice** plot; and the arrangement of several complete **lattice** plots together on a single page.

In the first case (the arrangement of panels and strips within a single plot), there are two useful arguments that can be specified in a call to a **lattice** plotting function: the `layout` argument and the `aspect` argument.

The `layout` argument consists of up to three values. The first two indicate the number of columns and rows of panels on each page and the third value indicates the number of pages. It is not necessary to specify all three values, as **lattice** provides sensible default values for any unspecified values. The following code produces a variation on Figure 4.5 by explicitly specifying that there should be a single column of three panels, via the `layout` argument, and that each panel must be "square," via the `aspect` argument. The final result is shown in Figure 4.7.

```
> xyplot(mpg ~ disp | factor(gear), data=mtcars,
         layout=c(1, 3), aspect=1)
```

The `aspect` argument specifies the aspect ratio (height divided by width) for the panels. The default value is `"fill"`, which means that panels expand to occupy as much space as possible. In the example above, the panels were all forced to be square by specifying `aspect=1`. This argument will also accept the special value `"xy"`, which means that the aspect ratio is calculated to satisfy the "banking to 45 degrees" rule proposed by Bill Cleveland.

As with the choice of colors and data symbols, a lot of work is done to select sensible default values for the arrangement of panels, so in many cases nothing special needs to be specified.

The problem of arranging multiple **lattice** plots on a page requires a different approach. A `trellis` object must be created (but not plotted) for each **lattice** plot, then the `print()` function is called, supplying arguments to specify the position of each plot. The following code demonstrates this idea by manually arranging three separate plots of automobile fuel efficiency for different numbers of gears in a column (see Figure 4.8).

Three **lattice** plots are produced and then positioned one above the other on a page. The `position` argument is used to specify their location, (`left`, `bottom`, `right`, `top`), as a proportion of the total page, and the `more` argument is used in the first and second `print()` calls to ensure that the second and third `print()` calls draw on the same page. Some extra work is done with the `xlim` and `ylim` arguments to make sure that the scales on the three plots match up.

```
> plot1 <- xyplot(mpg ~ disp, data=mtcars,
                  aspect=1, xlim=c(65, 480), ylim=c(9, 35),
                  subset=gear == 5)
> plot2 <- xyplot(mpg ~ disp, data=mtcars,
                  aspect=1, xlim=c(65, 480), ylim=c(9, 35),
                  subset=gear == 4)
> plot3 <- xyplot(mpg ~ disp, data=mtcars,
                  aspect=1, xlim=c(65, 480), ylim=c(9, 35),
                  subset=gear == 3)
> print(plot1, position=c(0, 2/3, 1, 1), more=TRUE)
> print(plot2, position=c(0, 1/3, 1, 2/3), more=TRUE)
> print(plot3, position=c(0, 0, 1, 1/3))
```

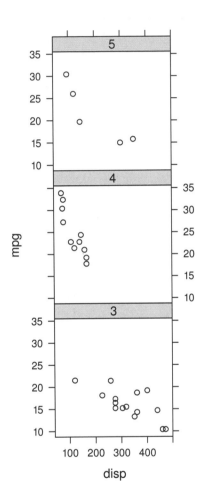

Figure 4.7
Controlling the layout of **lattice** panels. The **lattice** package arranges panels in a
sensible way by default, but there are several ways to force the panels to be arranged
in a particular layout. This figure shows a custom arrangement of the panels in the
plot from Figure 4.5.

Section 6.8 describes more flexible options for arranging multiple **lattice** plots, using the concepts and facilities of the **grid** system.

4.6 The `scales` argument and labeling axes

This section looks at controlling the scales and labeling of the axes in **lattice** plots.

The `scales` argument takes a list of different settings that influence the appearance of axes. The list can have sublists, named `x` and `y`, if the settings are intended to affect only the x-axes or only the y-axes.

In the following code, the `scales` argument is used to specify exactly where tick marks should appear on y-axes. This code also demonstrates that the `xlab` and `ylab` arguments can be expressions to allow the use of special formatting and special symbols. The plot produced by this code is shown in Figure 4.9.

```
> xyplot(mpg ~ disp | factor(gear), data=mtcars,
         layout=c(3, 1), aspect=1,
         scales=list(y=list(at=seq(10, 30, 10))),
         ylab="miles per gallon",
         xlab=expression(paste("displacement (in"^3, ")")))
```

Besides specifying the location and labels for tick marks, the `scales` argument can also be used to control the font used for tick labels (`font`), the rotation of the labels (`rot`), the range of values on the axes (`limits`), and whether these ranges should be the same for all panels (`relation="same"`) or allowed to vary between panels (`relation="free"`).

4.7 The `panel` argument and annotating plots

One advantage of the **lattice** graphics system is that it can produce extremely sophisticated plots from relatively simple expressions, especially with its multipanel conditioning feature. However, the cost of this is that the task of adding simple annotations of a **lattice** plot, such as adding extra lines or

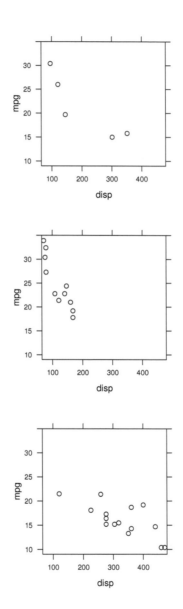

Figure 4.8
Arranging multiple **lattice** plots. This shows three separate **lattice** plots arranged together on a single page.

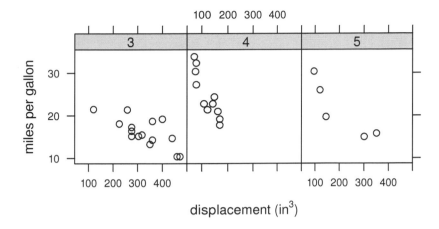

Figure 4.9
Modifying **lattice** axes. The placement of tick marks on the y-axis and the axis
labels have been customized in this plot.

text, is more complex compared to the same task in base graphics.

Extra drawing can be added to the panels of a **lattice** plot via the **panel**
argument. The value of this argument is a function, which gets called to draw
the contents of each panel.

The following code shows an example panel function. The main plot is once
again of the automobile fuel efficiency data, with three panels corresponding
to different numbers of gear. The panel function consists of calls to various predefined functions that are designed to add graphics to **lattice** panels. The first function call within the panel function is very important. The
panel.xyplot() function does the drawing that **xyplot()** would normally
have done if the **panel** argument had not been specified. In this case, it
draws a data symbol for each car. The other functions called in this panel
function are **panel.abline()** and **panel.text()**, which add a dashed horizontal line and a label to indicate an efficiency criterion of 29 miles per gallon.
The final result is shown in Figure 4.10.

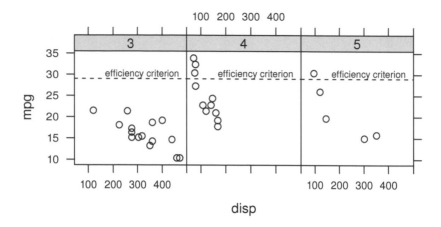

Figure 4.10
Adding annotations to **lattice** plots. The dashed horizontal lines and the labels
have been added to a standard **xyplot()** using a panel function.

```
> xyplot(mpg ~ disp | factor(gear), data=mtcars,
         layout=c(3, 1), aspect=1,
         panel=function(...) {
             panel.xyplot(...)
             panel.abline(h=29, lty="dashed")
             panel.text(470, 29.5, "efficiency criterion",
                        adj=c(1, 0), cex=.7)
         })
```

That panel function is a very simple one because it does exactly the same
thing in each panel. Things get more complicated if the panel function has to
produce different output for each panel. In that case, more attention has to
be paid to the arguments of the panel function.

In the simple example above, the panel function is defined with just an ellipsis
(...) argument. This means that any information that **lattice** sends to this
panel function is captured by the ellipsis argument and the panel function
simply passes the information on to **panel.xyplot()**.

Another common situation is that the extra graphics in a panel need to depend
on the x- and y-values that are plotted in that panel. The code below shows

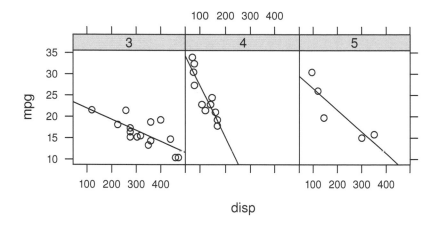

Figure 4.11
An example of a **lattice** panel function. A line of best fit has been added to each
panel in a standard `xyplot()` using a panel function.

an example, where the `panel.lmline()` function is called as part of the panel
function to draw a line of best fit to the data in each panel (see Figure 4.11).
The panel function now has explicit x- and y-arguments, which capture the
data values that **lattice** passes to each panel. These x- and y-values are passed
to `panel.lmline()` and to `panel.xyplot()` to produce the relevant output in
each panel. There is a lot of other information that **lattice** passes to the panel
function (see the argument list on the help page for `panel.xyplot()`), but
that is all simply passed through to `panel.xyplot()` via an ellipsis argument.

```
> xyplot(mpg ~ disp | factor(gear), data=mtcars,
        layout=c(3, 1), aspect=1,
        panel=function(x, y, ...) {
            panel.lmline(x, y)
            panel.xyplot(x, y, ...)
        })
```

As these examples have demonstrated, there are a number of predefined panel
functions available for adding output to a **lattice** panel, including both low-
level graphical primitives like points, and text and more high-level graphics
like grids and lines of best fit. For every high-level **lattice** plotting function

Table 4.2

A selection of predefined panel functions for adding graphical output to the panels of **lattice** plots.

Function	Description
panel.points()	Draw data symbols at locations (x, y)
panel.lines()	Draw lines between locations (x, y)
panel.segments()	Draw line segments between (x0, y0) and (x1, y1)
panel.arrows()	Draw line segments and arrowheads to the end(s)
panel.rect()	Draw rectangles with bottom-left corner at (xl, yl) and top-right corner at (xr, yt)
panel.polygon()	Draw one or more polygons with vertices (x, y)
panel.text()	Draw text at locations (x, y)
panel.abline()	Draw a line with intercept a and slope b
panel.curve()	Draw a function given by expr
panel.rug()	Draw axis ticks at x- or y-locations
panel.grid()	Draw a (gray) reference grid
panel.loess()	Draw a loess smooth through (x, y)
panel.violin()	Draw one or more violin plots
panel.smoothScatter()	Draw a smoothed 2D density of (x, y)

(see Table 4.1) there is also a corresponding default panel function, for example, `panel.xyplot()`, `panel.bwplot()`, and `panel.histogram()`. Table 4.2 provides a list of some other predefined panel functions.

One other important panel function is `panel.superpose()`, which is the default panel function whenever multiple groups are drawn within a panel (e.g., when the **group** argument is used). When writing a custom panel function for a **lattice** plot that has multiple groups in each panel, this function must be called to reproduce the default plotting behavior.

In addition to the **panel** argument for adding further drawing to **lattice** panels, there is a **strip** argument, which allows customization of the strips above each panel.

4.7.1 Adding output to a lattice plot

Unlike in the original Trellis implementation, it is also possible to add output to a complete **lattice** plot *after* the plot has been drawn (i.e., without using a panel function).

The function `trellis.focus()` can be used to return to a particular panel or strip of the current **lattice** plot in order to add further output using, for example, `panel.lines()` or `panel.points()`. The `trellis.unfocus()` function should be called after the extra drawing is complete. The function `trellis.panelArgs()` may be useful for retrieving the arguments (including the data) that were used to originally draw the panel.

Sections 6.8 and 7.14 show how **grid** provides more flexibility for navigating to different parts of a **lattice** plot and for adding further output.

4.8 `par.settings` and graphical parameters

An important feature of Trellis Graphics is the careful selection of default settings that are provided for many of the features of **lattice** plots. For example, the default data symbols and colors used to distinguish between different data series have been chosen so that it is easy to visually discriminate between them. Nevertheless, it is still sometimes desirable to be able to make alterations to the default settings for aspects like color and text size.

The examples at the start of this chapter demonstrated that many of the familiar standard arguments from base graphics, such as `col`, `lty`, and `lwd`, do the same job in **lattice** plots. These graphical parameters can also be set via a `par.settings` argument. For example, the original code for Figure 4.2, which draws both lines and point with custom `pch` and `lty` settings is reproduced below.

```
> xyplot(pressure ~ temperature, pressure,
         type="o", pch=16, lty="dashed",
         main="Vapor Pressure of Mercury")
```

The following code is an alternative way to produce the same result using the `par.settings` argument.

```
> xyplot(pressure ~ temperature, pressure,
         type="o",
         par.settings=list(plot.symbol=list(pch=16),
                           plot.line=list(lty="dashed")),
         main="Vapor Pressure of Mercury")
```

This approach works because **lattice** maintains a graphics state similar to the base graphics state: a large set of graphical parameter defaults.

The **lattice** graphical parameter settings consist of a large list of parameter groups and each parameter group is itself a list of parameter settings. These groups allow settings like color to be applied just to specific elements of a plot. For example, there is a `plot.line` parameter group consisting of `alpha`, `col`, `lty`, and `lwd` settings to control the color, line type, and line width for lines drawn between data locations. There is a separate `plot.symbol` group consisting of `alpha`, `cex`, `col`, `font`, `pch`, and `fill` settings to control the size, shape, and color of data symbols.

The settings in each parameter group affect some aspect of a **lattice** plot: some have a "global" effect, for example, the `fontsize` settings affect all text in a plot; some are more specific, for example, the `strip.background` setting affects the background color of strips; and some only affect a certain aspect of a certain sort of plot, for example, the `box.dot` settings affect only the dot that is plotted at the median value in boxplots.

The function `show.settings()` produces a picture representing some of the current graphical parameter settings. Figure 4.12 shows the settings for a black-and-white PostScript device.

The `par.settings` argument to high-level **lattice** plots allows specific graphical parameters to be set for a single plot, but, similar to `par()` in base graphics, the global default values can also be changed.

The current value of graphical parameter settings can be obtained using the `trellis.par.get()` function. For a list of all of the names of the parameter groups, type `names(trellis.par.get())`. If one of these group names is specified as the argument to `trellis.par.get()`, then only the relevant settings are returned. The following code shows how to obtain only the `add.text` group of settings.

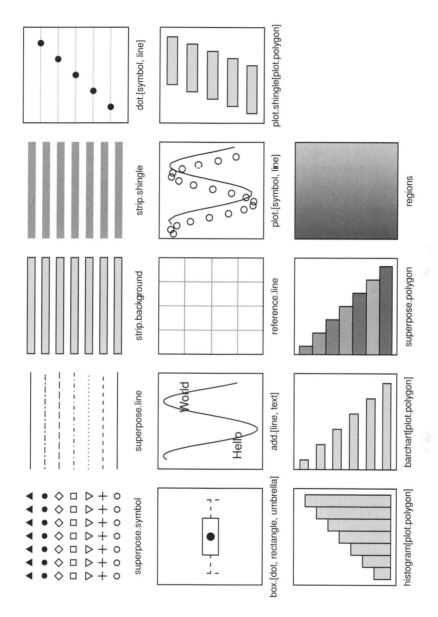

Figure 4.12

Some default **lattice** settings for a black-and-white PostScript device. This figure
was produced by the **lattice** function `show.settings()`.

```
> trellis.par.get("add.text")
```

```
$alpha
[1] 1
```

```
$cex
[1] 1
```

```
$col
[1] "#000000"
```

```
$font
[1] 1
```

```
$lineheight
[1] 1.2
```

The `trellis.par.set()` function can be used to specify new default values for graphical parameters. The value given to this function should be a list of lists. Only the components and groups that are to be changed need to be specified.

The following code demonstrates how to use `trellis.par.set()` to specify a new value for the `"col"` component of the `add.text` settings.

```
> trellis.par.set(list(add.text=list(col="red")))
```

A full set of **lattice** graphical parameter settings is called a *theme*. It is possible to specify such a theme and enforce a new "look and feel" for a plot, although choosing a complete set of defaults that all work together nicely is a difficult task. The **lattice** package currently provides one custom theme via the `col.whitebg()` function and there is a `simpleTheme()` function that makes creating a new theme easier.

There is much more that can be said about the **lattice** graphics system and there are many more plots that can be produced (see, for example, the **latticeExtra** package). However, the purpose of this chapter is just to enable us to produce a range of high-level plots in the **grid** graphics world. Chapters 6 and 7 will describe the tools within **grid** that can be used to customize, modify, and add to these **lattice** plots.

Chapter summary

The **lattice** package implements and extends the Trellis Graphics system for producing complete statistical plots. This system provides most standard plot types and a number of modern plot types with several important extensions. For a start, the layout and appearance of the plots is designed to maximize readability and comprehension of the information represented in the plot. Also, the system provides a feature called multipanel conditioning, which produces multiple panels of plots from a single data set, where each panel contains a different subset of the data. The **lattice** functions provide an extensive set of arguments for customizing the detailed appearance of a plot and there are functions that allow the user to add further output to a plot.

5

The Grammar of Graphics: The ggplot2 Package

Chapter preview

This chapter describes how to produce plots using the **ggplot2** package. There is a brief introduction to the concepts underlying the Grammar of Graphics paradigm as well as a description of the functions used to produce plots within this paradigm. The distinguishing feature of the **ggplot2** package is its ability to produce a very wide range of different plots from a relatively small set of fundamental components. Because **ggplot2** uses **grid** to draw plots, this chapter describes another way to produce a complete plot using the **grid** system.

The **ggplot2** package provides an interpretation and extension of the ideas in Leland Wilkinson's book *The Grammar of Graphics*. The **ggplot2** package represents a complete and coherent graphics system, completely separate from both base and **lattice** graphics.

The **ggplot2** package is built on **grid**, so it provides another way to generate complete plots within the **grid** world, but as with **lattice**, the package has so many features that it is unnecessary to encounter **grid** concepts for most applications.

The graphics functions that make up the graphics system are provided in an extension package called **ggplot2**. This package is not part of a standard R installation, so it must first be installed, then it can be loaded into R as follows.

```
> library(ggplot2)
```

This chapter presents a very brief introduction to **ggplot2**. Hadley Wickham's
book, *ggplot2: Elegant Graphics for Data Analysis*, provides much more detail
about the package.

5.1   Quick plots

For very simple plots, the qplot() function in **ggplot2** serves a similar purpose to the plot() function in base graphics. All that is required is to specify
the relevant data values and the qplot() function produces a complete plot.

For example, the following code produces a scatterplot of pressure versus
temperature using the **pressure** data set (see Figure 5.1).

```
> qplot(temperature, pressure, data=pressure)
```

This plot should be compared with Figures 1.1 and 4.1. The main differences
between this scatterplot and what is produced by the base graphics plot()
function, or **lattice**'s xyplot(), are just the default settings used for things
like the background grid, the plotting symbols, and the axis labeling.

There are also similarities in how the appearance of the plot can be modified.
For example, the following code adds a title to the plot using the argument
main.

```
> qplot(temperature, pressure, data=pressure,
        main="Vapor Pressure of Mercury")
```

However, **ggplot2** diverges quite rapidly from the other graphics systems if
further customizations are desired. For example, in order to plot both points
and lines on the plot, the following code is required (see Figure 5.2). Notice
that, like **lattice**, the **ggplot2** result has automatically resized the plot region
to provide room for the title.

```
> qplot(temperature, pressure, data=pressure,
        main="Vapor Pressure of Mercury",
        geom=c("point", "line"))
```

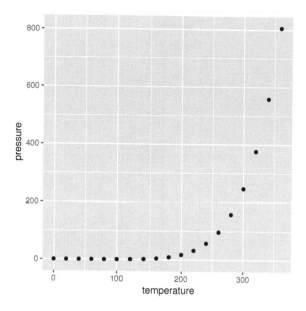

Figure 5.1

A scatterplot produced by the qplot() function from the **ggplot2** package. This plot is comparable to the base graphics plot in Figure 1.1.

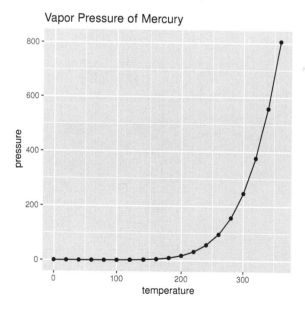

Figure 5.2

A scatterplot produced by the qplot() function from the **ggplot2** package, with a title and lines added. This plot is a modified version of 5.1.

In order to understand how this code works, rather than spending a lot of time on the `qplot()` function, it is useful to move on instead to the conceptual structure, the Grammar of Graphics, that underlies the **ggplot2** package.

5.2 The ggplot2 graphics model

The **ggplot2** package implements the Grammar of Graphics paradigm. This means that, rather than having lots of different functions, each of which produces a different sort of plot, there is a small set of functions, each of which produces a different sort of plot *component*, and those components can be *combined* in many different ways to produce a huge variety of plots.

The steps in creating a plot with **ggplot2** often come down to the following essentials:

- Define the data that you want to plot and create an empty plot object with `ggplot()`.

- Specify the graphics shapes, or *geoms*, that you are going to use to view the data (e.g., data symbols or lines) and add those to the plot with, for example, `geom_point()` or `geom_line()`.

- Specify which features, or *aesthetics*, of the shapes will be used to represent the data values (e.g., the `x`- and `y`-locations of data symbols) with the `aes()` function.

In summary, a plot is created by mapping data values via aesthetics to the features of geometric shapes (see Figure 5.3).

For example, to produce the simple plot in Figure 5.1, the data set is the `pressure` data frame, and the variables `temperature` and `pressure` are used as the `x` and `y` locations of data symbols. This is expressed by the following code.

```
> ggplot(pressure) +
      geom_point(aes(x=temperature, y=pressure))
```

A **ggplot2** plot is built up like this by creating plot components, or *layers*, and combining them using the `+` operator.

The following sections describe these ideas of geoms and aesthetics in more detail and go on to look at several other important components that allow for

Figure 5.3
A diagram showing how data is mapped to features of a geom (geometric shape) via aesthetics in **ggplot2**.

more complex plots that contain multiple groups, legends, facetting (similar to **lattice**'s multipanel conditioning), and more.

5.2.1 Why another graphics system?

Many of the plots that can be produced with **ggplot2** are very similar to the output of the base graphics system or the **lattice** graphics system, but there are several reasons for using **ggplot2** over the others:

- The default appearance of plots has been carefully chosen with visual perception in mind, like the defaults for **lattice** plots. The **ggplot2** style may be more appealing to some people than the **lattice** style.

- The arrangement of plot components and the inclusion of legends is automated. This is also like **lattice**, but the **ggplot2** facility is more comprehensive and sophisticated.

- Although the conceptual framework in **ggplot2** can take a little getting used to, once mastered, it provides a very powerful language for concisely expressing a wide variety of plots.

- The **ggplot2** package uses **grid** for rendering, which provides a lot of flexibility available for annotating, editing, and embedding **ggplot2** output (see Sections 6.9 and 7.15).

5.3 Data

The starting point for a plot is a set of data to visualize.

The examples throughout this section will make use of the `mtcars2` data set. This data set is based on the `mtcars` data set from the **datasets** package and contains information on 32 different car models, including the size of the car

engine (`disp`), its fuel efficiency (`mpg`), type of transmission (`trans`), number
of forward gears (`gear`), and number of cylinders (`cyl`). The first few lines of
the data set are shown below.

```
> head(mtcars2)
```

```
                  mpg cyl disp gear      trans
Mazda RX4         21.0  6  160    4     manual
Mazda RX4 Wag     21.0  6  160    4     manual
Datsun 710        22.8  4  108    4     manual
Hornet 4 Drive    21.4  6  258    3  automatic
Hornet Sportabout 18.7  8  360    3  automatic
Valiant           18.1  6  225    3  automatic
```

The following call to the `ggplot()` function creates a new plot for the `mtcars2`
data set. The data for a plot must always be a data frame.

```
> p <- ggplot(mtcars2)
```

The result of the `ggplot()` call is a `"ggplot"` object and, if we print this
object, a plot is drawn (see Figure 5.4).

```
> p
```

Our plot description contains no information yet about how to display the
data, so nothing is drawn. However, we will add more components to the plot
in later examples.

5.4   Geoms and aesthetics

The next step in creating a plot is to specify what sort of shape will be used
in the plot, for example, data symbols for a scatterplot or bars for a barplot.
This step also involves deciding which variables in the data set will be used
to control features of the shapes, for example, which variables will be used for
the (`x, y`) positions of the data symbols in a scatterplot.

The following code adds this information to the plot that was created in the
last section. This code produces a new `"ggplot"` object by adding information

Figure 5.4

A "ggplot" object that only contains data produces an empty plot.

that says to draw data symbols, using the `geom_point()` function, and that the `disp` variable should be used for the x location and the `mpg` variable should be used for the y location of the data symbols; these variables are mapped to the x and y aesthetics of the point geom, using the `aes()` function. The result is a scatterplot of fuel efficiency versus engine size (see Figure 5.5).

```
> p + geom_point(aes(x=disp, y=mpg))
```

Depending on what geom is being used to display the data, various other aesthetics are available. Another aesthetic that can be used with point geoms is the `shape` aesthetic. In the following code, the `gear` variable is associated with the data symbol shape so that cars with different numbers of forward gears are drawn with different data symbols (see Figure 5.5). Table 5.1 lists some of the common aesthetics for some common geoms.

```
> p + geom_point(aes(x=disp, y=mpg, shape=gear),
                 size=4)
```

This example also demonstrates the difference between *setting* an aesthetic and *mapping* an aesthetic. The `gear` variable is *mapped* to the `shape` aesthetic, using the `aes()` function, which means that the shapes of the data symbols are taken from the value of the variable and different data symbols will get different shapes. By contrast, the `size` aesthetic is *set* to the constant value of 4 (it is not part of the call to `aes()`), so all data symbols get this size.

The **ggplot2** package provides a range of geometric shapes that can be used to produce different sorts of plots. Other geoms include the standard graphical primitives, such as lines, text, and polygons, plus several more complex graphical shapes such as bars, contours, and boxplots (see later examples). Table 5.1 lists some of the common geoms that are available. As an example of a different sort of geom, the following code uses text labels rather than data symbols to plot the relationship between engine displacement and miles per gallon (see Figure 5.5). The locations of the text are the same as the locations of the data symbols from before, but the text drawn at each location is based on the value of the `gear` variable. This example also demonstrates another aesthetic, `label`, which is relevant for text geoms.

```
> p + geom_text(aes(x=disp, y=mpg, label=gear))
```

A plot can be made up of multiple geoms by simply adding further geoms to the plot description. The following code draws a plot consisting of both data symbols and a straight line that is based on a linear model fit to the data (see Figure 5.5). The line is defined by its `intercept` and `slope` aesthetics.

```
> lmcoef <- coef(lm(mpg ~ disp, mtcars2))
```

```
> p + geom_point(aes(x=disp, y=mpg)) +
      geom_abline(intercept=lmcoef[1], slope=lmcoef[2])
```

Specifying geoms and aesthetics provides the basis for creating a wide variety of plots with **ggplot2**. The remaining sections of this chapter introduce a number of other plot components within the **ggplot2** system, which are required to control the details of plots and which extend the range of plots even further.

5.5   Scales

Another important type of component that has not yet been mentioned is the *scale* component. In **ggplot2** this encompasses the ideas of both axes and legends on plots.

Scales have not been mentioned to this point because **ggplot2** will often automatically generate appropriate scales for plots. For example, the x-axes

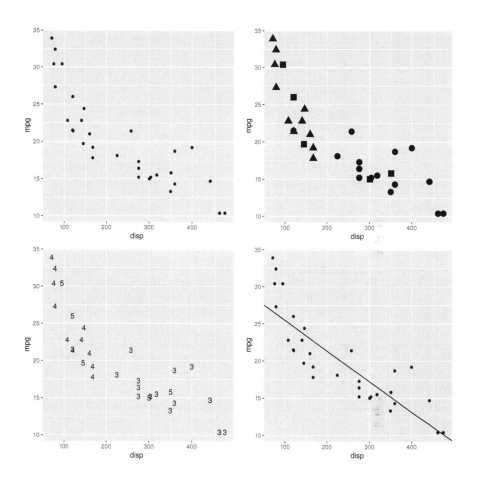

Figure 5.5

Variations on a scatterplot that shows the relationship between miles per gallon
(mpg) and engine displacement (disp): at top-left, a points geom is used to plot
data symbols; at top-right, the **shape** aesthetic of the points geom is used to plot
different data symbols for cars with different numbers of forward gears; at bottom-
left, a text geom is used to plot labels rather than data symbols; and at bottom-right,
both a points geom and an abline geom are used on the same plot to draw both
data symbols and a straight line (of best fit).

Table 5.1

Some of the common geoms and their common aesthetics that are available in the **ggplot2** graphics system. Many geoms have `color`, `size`, and `group` aesthetics. The `size` aesthetic means size of shape for points, height for text, and width for lines and it is in units of millimeters.

Geom	Description	Aesthetics
geom_point()	Data symbols	x, y, shape, fill
geom_line()	Line (ordered on x)	x, y, linetype
geom_path()	Line (original order)	x, y, linetype
geom_text()	Text labels	x, y, label, angle, hjust, vjust
geom_rect()	Rectangles	xmin, xmax, ymin, ymax, fill, linetype
geom_polygon()	Polygons	x, y, fill, linetype
geom_segment()	Line segments	x, y, xend, yend, linetype
geom_bar()	Bars	x, fill, linetype, weight
geom_histogram()	Histogram	x, fill, linetype, weight
geom_boxplot()	Boxplots	x, y, fill, weight
geom_density()	Density	x, y, fill, linetype
geom_contour()	Contour lines	x, y, fill, linetype
geom_smooth()	Smoothed line	x, y, fill, linetype
Common to many geoms		color, size, group

and y-axes on the previous plots in this section are actually scale components that have been automatically generated by **ggplot2**.

One reason for explicitly adding a scale component to a plot is to override the detail of the scale that **ggplot2** creates. For example, the following code explicitly sets the axis labels using the `scale_x_continuous()` and `scale_y_continuous()` functions (see Figure 5.6).

```
> p + geom_point(aes(x=disp, y=mpg)) +
    scale_y_continuous(name="miles per gallon") +
    scale_x_continuous(name="displacement (cu.in.)")
```

It is also possible to control features such as the limits of the axis, where the tick marks should go, and what the tick labels should look like. Table 5.2 shows some of the common scale functions and their arguments. In the following code, the limits of the y-axis are widened to include zero (see Figure 5.6).

```
> p + geom_point(aes(x=disp, y=mpg)) +
    scale_y_continuous(limits=c(0, 40))
```

The **ggplot2** package also automatically creates legends when it is appropriate to do so. For example, in the following code, the `color` aesthetic is mapped to the `trans` variable in the `mtcars` data frame, so that the data symbols are colored according to what sort of transmission a car has. This automatically produces a legend to display the mapping between type of transmission and color.

```
> p + geom_point(aes(x=disp, y=mpg,
                      color=trans), size=4)
```

The plot resulting from the above code is not shown because this example demonstrates another important role that scales play in the **ggplot2** system.

When the `aes()` function is used to set up a mapping, the values of a variable are used to generate values of an aesthetic. Sometimes this is very straightforward. For example, when the variable `disp` is mapped to the aesthetic `x` for a points geom, the numeric values of `disp` are used directly as `x` locations for the points.

However, in other cases, the mapping is less obvious. For example, when the variable `trans`, with values `"manual"` and `"automatic"`, is mapped to the aesthetic `color` for a points geom, what color does the value `"manual"` correspond to?

Table 5.2
Some of the common scales that are available in the **ggplot2** graphics system. Most scales have `name`, `breaks`, `labels`, `limits` parameters. For every x-axis scale there is a corresponding y-axis scale.

Scale	Description	Parameters
`scale_x_continuous()`	Continuous axis	`expand`, `trans`
`scale_x_discrete()`	Categorical axis	
`scale_x_date()`	Date axis	`major`, `minor`, `format`
`scale_shape()`	Symbol shape legend	
`scale_linetype()`	Line pattern legend	
`scale_color_manual()`	Symbol/line color legend	`values`
`scale_fill_manual()`	Symbol/bar fill legend	`values`
`scale_size()`	Symbol size legend	`trans`, `to`
Common to most scales		`name`, `breaks`, `labels`, `limits`

As usual, **ggplot2** provides a reasonable answer to this question by default, but a second reason for explicitly adding a scale component to a plot is to explicitly control this mapping of variable values to aesthetic values (see Figure 5.7). For example, the following code uses the `scale_color_manual()` function to specify the two colors (shades of gray) that will correspond to the two values of the `trans` variable (see Figure 5.6).

```
> p + geom_point(aes(x=disp, y=mpg,
                      color=trans), size=4) +
      scale_color_manual(values=c(automatic=gray(2/3),
                                  manual=gray(1/3)))
```

5.6   Statistical transformations

In the examples so far, data values have been mapped directly to aesthetic settings. For example, the numeric `disp` values have been used as x-locations for data symbols and the levels of the `trans` factor have been associated with different symbol colors.

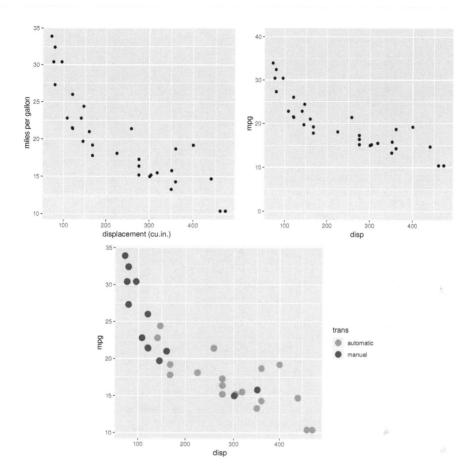

Figure 5.6

Scatterplots that have explicit scale components to control the labeling of axes or the mapping from variable values to colors: at top-left, the x-axis and y-axis labels are specified explicitly; at top-right, the y-axis range has been expanded; and the bottom plot has an explicit mapping between transmission type and shades of gray.

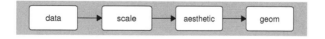

Figure 5.7

A diagram showing how the mapping of data to the features of geometric shapes is controlled by a scale. The scale specifies how data values are mapped to aesthetic values.

Figure 5.8
A diagram showing how the scaled data may undergo a statistical transformation
before being mapped to the values of an aesthetic.

Some geoms do not use the raw data values like this. Instead, the data values
undergo some form of statistical transformation, or *stat*, and the transformed
values are mapped to aesthetics (see Figure 5.8).

A good example of this sort of thing is the bar geom. This geom counts
the number of times each different data value occurs and uses the counts as
the data to plot. For example, in the following code, the `trans` variable is
mapped to the x aesthetic in the `geom_bar()` call. This establishes that the
x-locations of the bars should be the levels of `trans`, but heights of the bars
(the y aesthetic) is automatically generated from the counts of each level of
`trans` to produce a bar plot (see Figure 5.9).

```
> p + geom_bar(aes(x=trans))
```

The stat that is used in this case is a `"count"` stat. Another option is an
identity stat, which does not transform the data at all. The following code
shows how to explicitly set the stat for a geom by creating the same bar plot
from data that have already been counted.

```
> transCounts <- as.data.frame(table(mtcars2$trans))
> transCounts
```

```
      Var1 Freq
1 automatic   19
2    manual   13
```

Now, both the x and the y aesthetics are set explicitly for the bar geom and
the stat is set to `"identity"` to tell the geom not to count again. The result
of this code is exactly the same as the left plot in Figure 5.9.

```
> ggplot(transCounts) +
      geom_bar(aes(x=Var1, y=Freq), stat="identity")
```

The following code presents another common transformation, which involves
smoothing the original values. In this code, a `smooth` geom is added to the

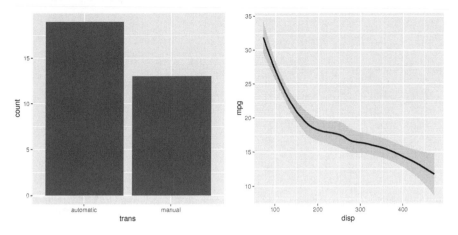

Figure 5.9
Examples of geoms with stat components: a bar geom, which uses a binning stat, and a smooth geom, which uses a smoother stat.

original empty plot. Rather than drawing a line through the original (x, y) values, this geom draws a smoothed line (plus a confidence band; see Figure 5.9).

```
> p + geom_smooth(aes(x=disp, y=mpg))
```

A similar result (without the confidence band) can be obtained using a line geom and explicitly specifying a "smooth" stat, as shown below.

```
> p + geom_line(aes(x=disp, y=mpg), stat="smooth")
```

Yet another alternative is to add an explicit stat component, as in the following code. This works because stat components automatically have a geom associated with them, just as geoms automatically have a stat associated with them. The default geom for a smoother stat is a line. The result of this code is exactly the same as the right plot in Figure 5.9.

```
> p + stat_smooth(aes(x=disp, y=mpg))
```

Similarly, the bar plot in Figure 5.9 could be created with an explicit count stat component, as shown below. The default geom for a count stat is a bar.

```
> p + stat_count(aes(x=trans))
```

Table 5.3
Some of the common stats that are available in the **ggplot2** graphics system.

Stat	Description	Parameters
stat_identity()	No transformation	–
stat_count()	Counts	–
stat_bin()	Binning	binwidth, origin
stat_smooth()	Smoother	method, se, n
stat_boxplot()	Boxplot statistics	width
stat_contour()	Contours	breaks

One advantage of this approach is that parameters of the stat, such as the smoothing method for a smooth stat or the binwidths for binning data, can be specified clearly as part of the stat. For example, the following code controls the method for the smooth stat to get a straight line (the result is similar to the line in Figure 5.5).

```
> p + stat_smooth(aes(x=disp, y=mpg), method="lm")
```

Table 5.3 shows some common **ggplot2** stats and their parameters.

5.7   The group aesthetic

Previous examples have demonstrated that **ggplot2** automatically handles plotting multiple groups of data on a plot. For example, in the following code, by introducing the trans variable as an aesthetic that controls shape, two groups of data symbols are generated on the plot and a legend is produced (the scale_shape_manual() function is used to control the mapping from trans to data symbol shape; see Figure 5.10).

```
> p + geom_point(aes(x=disp, y=mpg, shape=trans)) +
      scale_shape_manual(values=c(1, 3))
```

It is also useful to be able to explicitly force a grouping for a plot and this can be achieved via the group aesthetic. For example, the following code adds a smoother stat to a scatterplot where the data symbols are all the same, but there are separate smoothed lines for separate types of transmissions; the

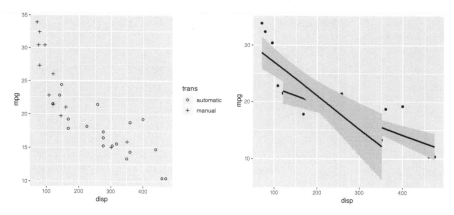

Figure 5.10
The **group** aesthetic in **ggplot2**. At left, mapping the **shape** aesthetic for point
geoms automatically generates a legend. At right, mapping the **group** aesthetic for
a smoother stat generates separate smoothed lines for different groups.

group aesthetic is set for the smoother stat. The **method** parameter is also
set for the smoother stat so that the result is a straight line of best fit (see
Figure 5.10).

```
> ggplot(mtcars2, aes(x=disp, y=mpg)) +
     geom_point() +
     stat_smooth(aes(group=trans),
                 method="lm")
```

Notice that in the code above, aesthetic mappings have been specified in the
call to **ggplot()**. This is more efficient when several components in a plot
share the same aesthetic settings.

5.8   Position adjustments

Another detail that **ggplot2** often handles automatically is the problem of
how to arrange geoms that overlap with each other. For example, the following
code produces a bar plot of the number of cars with different transmissions,
but also with the number of cylinders, **cyl**, mapped to the fill color for the
bars (see Figure 5.11). The **color** aesthetic for the bars is set to **"black"** to

provide borders for the bars and the fill color scale is explicitly set to three
shades of gray.

```
> p + geom_bar(aes(x=trans, fill=factor(cyl)),
              color="black") +
      scale_fill_manual(values=gray(1:3/3))
```

There are three bars in this plot for automatic transmission cars (i.e., three
bars share the same x-location). Rather than draw these bars over the top of
each other, **ggplot2** has automatically stacked them up. This is an example
of *position adjustment*.

An alternative is to use a "dodge" position adjustment, which places the bars
side-by-side. This is shown in the following code and the result is shown in
Figure 5.11.

```
> p + geom_bar(aes(x=trans, fill=factor(cyl)),
              color="black",
              position="dodge") +
      scale_fill_manual(values=gray(1:3/3))
```

Another option is a "fill" position adjustment. This expands the bars to fill
the available space to produce a spine plot (see Figure 5.11).

```
> p + geom_bar(aes(x=trans, fill=factor(cyl)),
              color="black",
              position="fill") +
      scale_fill_manual(values=gray(1:3/3))
```

5.9   Coordinate transformations

Section 5.5 described how scale components can be used to control the mapping between data values and the values of an aesthetic (e.g., map the `trans`
value "automatic" to the `color` value `gray(2/3)`).

Another way to view this feature is as a *transformation* of the data values into
the aesthetic domain. Another example of a transformation of data values
is to use log axes on a plot. The following code does this for the plot of
engine displacement versus miles per gallon via the `trans` argument of the
`scale_x_continuous()` function. The result is shown in Figure 5.12.

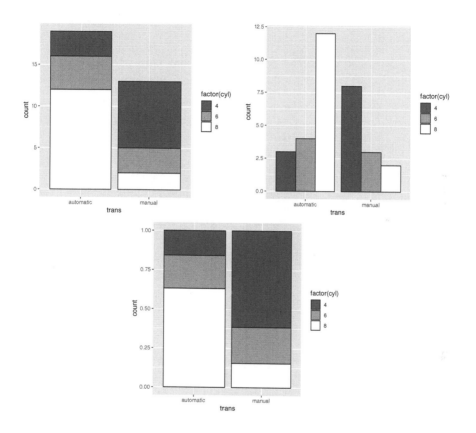

Figure 5.11

Examples of position adjustments in **ggplot2**: at top-left, the bars are `"stacked"`; at top-right, the bar position is `"dodge"` so the bars are side-by-side; and at the bottom, the position is `"fill"`, so the bars are scaled to fill the available (vertical) space.

```
> p + geom_point(aes(x=disp, y=mpg)) +
      scale_y_continuous(trans="log",
                            breaks=seq(10, 40, 10)) +
      scale_x_continuous(trans="log",
                            breaks=seq(100, 400, 100)) +
      geom_line(aes(x=disp, y=mpg), stat="smooth",
                method="lm")
```

This is another reason for using an explicit scale component in a plot. Notice that the data are transformed by the scale *before* any stat components are applied (see Figure 5.8), so the line is fitted to the log transformed data.

Another type of transformation is also possible in **ggplot2**. There is a coordinate system component, or *coord*, which by default is simple linear cartesian coordinates, but this can be explicitly set to something else.

For example, the following code adds a coordinate system component to the previous plot, using the `coord_trans()` function. This transformation says that both dimensions should be exponential.

```
> p + geom_point(aes(x=disp, y=mpg)) +
      scale_x_continuous(trans="log") +
      scale_y_continuous(trans="log") +
      geom_line(aes(x=disp, y=mpg), stat="smooth",
                method="lm") +
      coord_trans(x="exp", y="exp")
```

This sort of transformation occurs *after* the plot geoms have been created and controls how the graphical shapes are drawn on the page or screen (see Figure 5.13). In this case, the effect is to reverse the transformation of the data, so that the data points are back in their familiar arrangement and the line of best fit, which was fitted to the logged data, has become a curve (see Figure 5.12).

Another example of a coordinate system in **ggplot2** is polar coordinates, where the x- and y-values are treated as angle and radius values. The following code creates a normal, cartesian coordinate system, stacked barplot showing the number of cars with automatic versus manual transmissions (see Figure 5.12).

```
> p + geom_bar(aes(x="", fill=trans)) +
      scale_fill_manual(values=gray(1:2/3))
```

This next code sets the coordinate system to be polar, so that the y-values (the heights of the bars) are treated as angles and x-values (the width of the bar) is a (constant) radius. The result is a pie chart (see Figure 5.12).

```
> p + geom_bar(aes(x="", fill=trans)) +
      scale_fill_manual(values=gray(1:2/3)) +
      coord_polar(theta="y")
```

5.10   Facets

Facetting means breaking the data into several subsets and producing a separate plot for each subset on a single page. This is similar to **lattice**'s idea of multipanel conditioning and is also known as producing *small multiples*

The `facet_wrap()` function can be used to add facetting to a plot. The main argument to this function is a formula that describes the variable to use for subsetting the data. For example, in the following code a separate scatterplot is produced for each value of **gear** (see Figure 5.14). The **nrow** argument is used here to ensure a single row of plots is produced.

```
> p + geom_point(aes(x=disp, y=mpg)) +
      facet_wrap(~ gear, nrow=1)
```

There is also a `facet_grid()` function for producing plots arranged on a grid. The main difference is that the formula argument is of the form **y ~ x** and a separate row of plots is produced for each level of **y** and a separate column of plots is produced for each level of **x**.

5.11   Themes

The **ggplot2** package takes a different approach to controlling the appearance of graphical objects, by separating output into data and non-data elements. Geoms represent the data-related elements of a plot and aesthetics are used to control the appearance of a geom, as was described in Section 5.4. This section looks at how to control the non-data elements of a plot, such as the labels and lines used to create the axes and legends.

The collection of graphical parameters that control non-data elements is called a *theme* in **ggplot2**. A theme can be added as another component to a plot

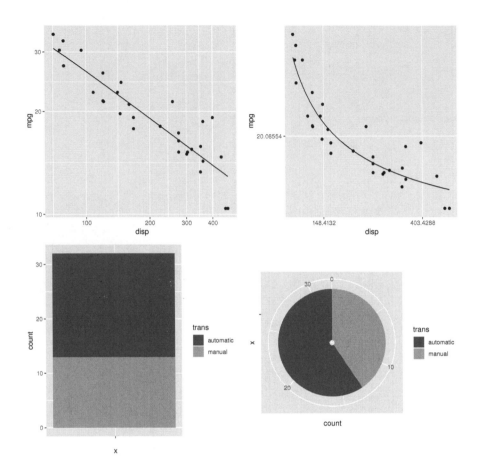

Figure 5.12
Examples of coordinate system transformations in **ggplot2**: at top-left is a cartesian
plot of logged data with linear axes; at top-right is a cartesian plot of logged data
with exponential axes; at bottom-left is a cartesian stacked barplot; and at bottom-
right is a polar stacked barplot (a pie chart).

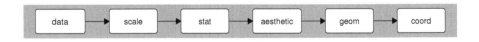

Figure 5.13
A diagram showing how geometric shapes may be transformed by a coordinate
system before they are drawn on the page or screen.

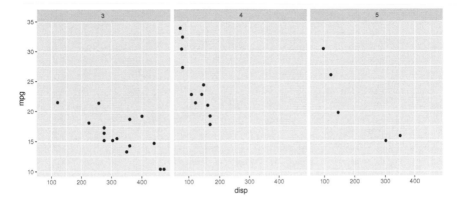

Figure 5.14
A facetted **ggplot2** scatterplot. A separate panel is produced for each level of a facetting variable, **gear**.

in the now-familiar way. For example, the following code creates a basic scatterplot, but changes the basic color settings for the plot using the function `theme_bw()`. Instead of the standard gray background with white grid lines, this plot has a white background with gray gridlines (see Figure 5.15).

```
> p + geom_point(aes(x=disp, y=mpg)) +
      theme_bw()
```

It is also possible to set just specific *theme elements* of the overall theme for a plot. This requires the `theme()` function and one of the *element functions* to specify the new setting. For example, the following code uses the `element_text()` function to make the y-axis label horizontal (see Figure 5.15). This example sets the text angle of rotation (and the vertical justification); it is also possible to set other parameters such as text font, color, and horizontal justification.

```
> p + geom_point(aes(x=disp, y=mpg)) +
      theme(axis.title.y=element_text(angle=0, vjust=.5))
```

There are other functions for setting graphical parameters for lines, segments, and rectangles, plus `element_blank()`, which removes the relevant plot element completely (see Figure 5.15).

```
> p + geom_point(aes(x=disp, y=mpg)) +
      theme(axis.title.y=element_blank())
```

Table 5.4

Some of the common plot elements in the **ggplot2** graphics system.
The type implies which element function should be used to provide
graphical parameter settings (e.g., text implies `element_text()`).

Element	Type	Description
`axis.text.x`	text	X-axis tick labels
`legend.text`	text	Legend labels
`panel.background`	rect	Background of panel
`panel.grid.major`	line	Major grid lines
`panel.grid.minor`	line	Minor grid lines
`plot.title`	text	Plot title
`strip.background`	rect	Background of facet labels
`strip.text.x`	text	Text for horizontal strips

Table 5.4 shows some of the plot elements that can be controlled in this way.

The `labs()` function can be used to control the labelling of the plot. For
example, the following code specifies an overall title for a scatterplot (see
Figure 5.15).

```
> p + geom_point(aes(x=disp, y=mpg)) +
      labs(title="Vehicle Fuel Efficiency")
```

5.12   Annotating

With the emphasis on mapping values from a data frame to aesthetics of
geoms, it may not be immediately obvious how to create custom annotations
on a plot with **ggplot2**.

One approach is just to make use of the ability to *set* aesthetics rather than
mapping them. For example, the following code shows how to add a single
horizontal line to a scatterplot by setting the `yintercept` aesthetic of an
`hline` geom to a specific value. The result is shown in Figure 5.16.

```
> p + geom_point(aes(x=disp, y=mpg)) +
      geom_hline(yintercept=29)
```

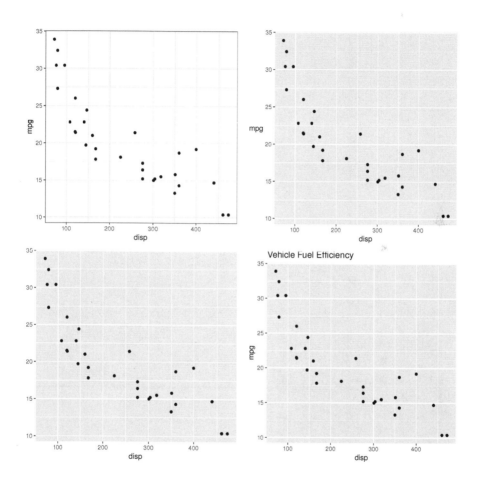

Figure 5.15
Some examples of themes in **ggplot2**: at top-left, the overall default style has been set to `theme_bw`; at top-right, the y-axis label has been rotated to horizontal; at bottom-left, the y-axis label has been removed altogether; at bottom-right, the plot has been given an overall title.

Another option is to make use of the fact that the functions that create geoms are actually creating a complete *layer*, just with many components of the layer either inheriting or automatically generating default values. In particular, a geom inherits its data source from the original "ggplot" object that forms the basis for the plot. However, it is possible to specify a new data source for a geom instead.

In order to demonstrate this idea, the following code generates a data frame containing various fuel efficiency (lower) limits for different classes of vehicle. These come from Criterion 4 of the Green Communities Grant Program, which is run by the Massachusetts Department of Energy Resources.[*]

```
> gcLimits <-
      data.frame(category=c("2WD car",
                    "4WD car",
                    "2WD small pick-up truck",
                    "4WD small pick-up truck",
                    "2WD std pick-up truck",
                    "4WD std pick-up truck"),
                 limit=c(29, 24, 20, 18, 17, 16))
```

The following code creates a scatterplot from the mtcars2 data set and adds some extra lines and text based on this new gcLimits data set. The data argument to the geom functions is used to explicitly specify the data source for these geoms, so the aesthetic mappings for these geoms make use of variables from the gcLimits data frame rather than the mtcars2 data frame. The final result is shown in Figure 5.16.

```
> p + geom_point(aes(x=disp, y=mpg)) +
      geom_hline(data=gcLimits,
                  aes(yintercept=limit),
                  linetype="dotted") +
      geom_text(data=gcLimits,
                  aes(y=limit + .1, label=category),
                  x=70, hjust=0, vjust=0, size=3)
```

---

[*]https://www.mass.gov/orgs/massachusetts-department-of-energy-resources

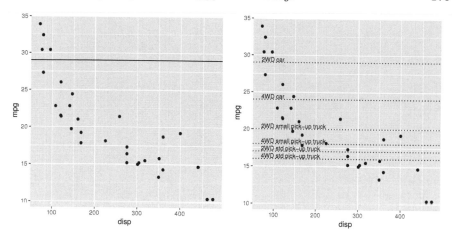

Figure 5.16

Some examples of annotation in **ggplot2**: at left, a single horizontal line has been added by setting a geom aesthetic (rather than mapping the aesthetic) and, at right, several horizontal lines and text labels have been added by using a completely new data set for the relevant geoms.

5.13 Extending ggplot2

Because **ggplot2** is based on a set of plot components that are combined to form plots, developing a new type of plot is usually simply a matter of combining the existing components in a new way.

Hadley Wickham's **ggplot2** book provides further discussion, including advice on how to write a high-level function for producing a plot from **ggplot2** functions.

Chapter summary

The **ggplot2** package implements and extends the Grammar of Graphics paradigm for statistical plots. The qplot() function works like plot() in very simple cases. Otherwise, a plot is created from basic components: a data frame, plus a set of geometric shapes (geoms), with a set of mappings from data values to properties of the shapes (aesthetics). Legends and axes are generated automatically, but the detailed appearance of all aspects of a plot can still be controlled. Facetted (multipanel) plots are also possible.

6

The grid Graphics Model

Chapter preview

This chapter describes the fundamental tools that **grid** provides for drawing graphical scenes. There are basic features such as functions for drawing lines, rectangles, and text, together with more sophisticated and powerful concepts such as viewports, layouts, and units, which allow basic output to be located and sized in very flexible ways.

This chapter is useful for drawing a wide variety of pictures, including statistical plots from scratch, and it is useful for adding to or modifying plots created by **lattice** or **ggplot2**.

The functions that make up the **grid** graphics system are provided in an extension package called **grid**. The **grid** system is loaded into R as follows.

```
> library(grid)
```

The **grid** graphics system only provides low-level graphics functions. There are no high-level functions for producing complete plots. Section 6.1 briefly introduces the concepts underlying the **grid** system, but this only provides an indication of how to work with **grid** and some of the things that are possible. An effective direct use of **grid** functions requires a deeper understanding of the **grid** system (see later sections of this chapter and Chapter 7).

The **lattice** and **ggplot2** packages described in Chapters 4 and 5 provide extensive demonstrations of the high-level results that can be achieved using **grid**. Other examples in this book are Figures 1.9, 1.10, 1.12, and 1.13 in Chapter 1.

6.1   A brief overview of grid graphics

This chapter describes how to use **grid** to produce graphical output. There
are functions to produce basic output, such as lines and rectangles and text,
and there are functions to establish the context for drawing, such as specifying
where output should be placed and what colors and fonts to use for drawing.

Like the base system, **grid** follows the painters model, with later output obscuring any earlier output that it overlaps. In this way, images can be constructed incrementally using **grid** by calling functions in sequence to add more
and more output.

There are **grid** functions to draw primitive graphical output such as lines,
text, and polygons, plus some slightly higher-level graphical components such
as axes (see Section 6.2). Complex graphical output is produced by making a
sequence of calls to these primitive functions.

The colors, line types, fonts, and other aspects that affect the appearance of
graphical output are controlled via a set of graphical parameters (see Section
6.4).

The **grid** system provides no predefined regions for graphical output, but there
is a powerful facility for defining regions, based on the idea of a *viewport* (see
Section 6.5). It is quite simple to create a set of regions that are convenient
for producing a single plot (see the example in the next section), but it is also
possible to produce very complex sets of regions such as those used in the
production of Trellis plots (see Chapter 4).

All viewports have a large set of coordinate systems associated with them
so that it is possible to position and size output in physical terms (e.g., in
centimeters) as well as relative to the scales on axes, and in a variety of other
ways (see Section 6.3).

All **grid** output occurs relative to the current viewport (region) on a page. In
order to start a new page of output, the user must call the `grid.newpage()`
function.

In addition to the side effect of producing graphical output, **grid** graphics
functions produce objects representing output. These objects can be saved
to produce a persistent record of a plot, and other **grid** functions exist to
modify these graphical objects. For example, it is possible to query an object
to determine its width on the page so that other drawing can be placed relative to the position of that object. It is also possible to work entirely with
graphical descriptions, without producing any output. Functions for working

with graphical objects are described in detail in Chapter 7.

6.1.1   A simple example

The following example demonstrates the construction of a simple scatterplot using **grid**. This is more work than a single function call to produce the plot, but it shows some of the advantages that can be gained by producing the plot using **grid**.

This example uses the **pressure** data to produce a scatterplot much like that in Figure 1.1.

Firstly, some regions are created that will correspond to the "plot region" (the area within which the data symbols will be drawn) and the "margins" (the area used to draw axes and labels).

The following code creates two viewports. The first viewport is a rectangular region that leaves space for five lines of text at the bottom, four lines of text at the left side, two lines at the top, and two lines to the right. The second viewport is in the same location as the first, but it has x- and y-scales corresponding to the range of the pressure data to be plotted.

```
> pushViewport(plotViewport(c(5, 4, 2, 2)))
> pushViewport(dataViewport(pressure$temperature,
                            pressure$pressure,
                            name="plotRegion"))
```

The following code draws the scatterplot one piece at a time. The output from **grid** functions is drawn relative to the most recent viewport, which in this case is the viewport with the appropriate axis scales. The data symbols are drawn relative to the x- and y-scales, a rectangle is drawn around the entire plot region, and x- and y-axes are drawn to represent the scales.

```
> grid.points(pressure$temperature, pressure$pressure,
              name="dataSymbols")
> grid.rect()
> grid.xaxis()
> grid.yaxis()
```

Adding labels to the axes demonstrates the use of the different coordinate systems that are available in **grid**. The label text is drawn outside the edges of the plot region and is positioned in terms of a number of lines of text (i.e., the height that a line of text would occupy).

```
> grid.text("temperature", y=unit(-3, "line"))
> grid.text("pressure", x=unit(-3, "line"), rot=90)
```

The obvious result of running the above code is the graphical output (see the top-left image in Figure 6.1). Less obvious is the fact that several objects have been created. There are objects representing the viewport regions and there are objects representing the graphical output. The following code makes use of this fact to modify the plotting symbol from a circle to a triangle (see the top-right image in Figure 6.1). The object representing the data symbols was named "dataSymbols" (see the code above) and this name is used to find that object and modify it using the grid.edit() function.

```
> grid.edit("dataSymbols", pch=2)
```

The next piece of code makes use of the objects representing the viewports. The upViewport() and downViewport() functions are used to navigate between the different viewport regions to perform some extra annotations. First of all, a call to the upViewport() function is used to go back to working within the entire page so that a dashed rectangle can be drawn around the complete plot.

```
> upViewport(2)
> grid.rect(gp=gpar(lty="dashed"))
```

Next, the downViewport() function is used to return to the plot region to add a text annotation that is positioned relative to the scale on the axes of the plot (see bottom-right image in Figure 6.1).

```
> downViewport("plotRegion")
> grid.text("Pressure (mm Hg)\nversus\nTemperature (Celsius)",
            x=unit(150, "native"), y=unit(600, "native"))
```

The final scatterplot is still quite simple in this example, but the techniques that were used to produce it are very general and powerful. It is possible to produce a very complex plot, yet still have complete access to modify and add to any part of the plot.

In the remaining sections of this chapter, the basic **grid** concepts of viewports and units are discussed in full detail. A complete understanding of the **grid** system will be useful in two ways: it will allow the user to produce very complex images from scratch and it will allow the user to work effectively with complex **grid** output that is produced by other people's code, for example plots that are produced using **lattice** or **ggplot2**.

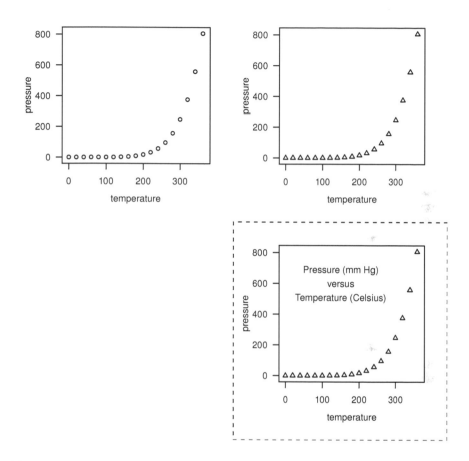

Figure 6.1

A simple scatterplot produced using **grid**. The top-left plot was constructed from a series of calls to primitive **grid** functions that produce graphical output. The top-right plot shows the result of calling the `grid.edit()` function to interactively modify the plotting symbol. The bottom-right plot was created by making calls to `upViewport()` and `downViewport()` to navigate between different drawing regions and adding further output (a dashed border and text within the plot).

6.2   Graphical primitives

The most simple **grid** functions to understand are those that draw something. There are a set of **grid** functions for producing basic graphical output such as lines, circles, and text.* Table 6.1 lists the full set of these functions.

The first arguments to most of these functions are a set of locations and dimensions for the graphical object to draw. For example, `grid.rect()` has arguments `x`, `y`, `width`, and `height` for specifying the locations and sizes of the rectangles to draw. An important exception is the `grid.text()` function, which requires the text to draw as its first argument. The text to draw may be a character vector or an R expression (to produce special symbols and formatting; see Section 10.5).

In most cases, multiple locations and sizes can be specified and multiple primitives will be produced in response. For example, the following function call produces 100 circles because 100 locations and radii are specified (see Figure 6.2).

```
> grid.circle(x=seq(0.1, 0.9, length=100),
              y=0.5 + 0.4*sin(seq(0, 2*pi, length=100)),
              r=abs(0.1*cos(seq(0, 2*pi, length=100))))
```

The `grid.move.to()` and `grid.line.to()` functions are unusual in that they both only accept one location. These functions refer to and modify a "current location." The `grid.move.to()` function sets the current location and `grid.line.to()` draws from the current location to a new location, then sets the current location to be the new location. The current location is not used by the other drawing functions. In most cases, `grid.lines()` will be more convenient, but `grid.move.to()` and `grid.line.to()` are useful for drawing lines across multiple viewports (also see Section 6.5.1).

The difference between `grid.lines()` and `grid.polyline()` is that the latter has an `id` argument. That argument can be used to split the (`x`, `y`) locations into separate lines.

The `grid.curve()` function draws a curve between two locations, which is useful in drawing simple diagrams. Several arguments control the shape of

---

*All of these functions are of the form `grid.*()` and, for each one, there is a corresponding `*Grob()` function that creates an object containing a description of primitive graphical output, but does not draw anything. The `*Grob()` versions are addressed fully in Chapter 7.

Table 6.1

Graphical primitives in **grid**. This is the complete set of low-level functions that produce graphical output. For each function that produces graphical output (leftmost column), there is a corresponding function that returns a graphical object containing a description of graphical output instead of producing graphical output (right-most column). The latter set of functions is described further in Chapter 7.

Function to Produce Output	Description	Function to Produce Object
grid.move.to()	Set the current location.	moveToGrob()
grid.line.to()	Draw a line from the current location to a new location and reset the current location.	lineToGrob()
grid.lines()	Draw a single line through multiple locations in sequence.	linesGrob()
grid.polyline()	Draw multiple lines through multiple locations in sequence.	polylineGrob()
grid.segments()	Draw multiple lines between pairs of locations.	segmentsGrob()
grid.xspline()	Draw smooth curve relative to control points.	xsplineGrob()
grid.bezier	Draw an (approximate) Bezier curve.	bezierGrob
grid.rect()	Draw rectangles given locations and sizes.	rectGrob()
grid.roundrect()	Draw rectangles with rounded corners, given locations and sizes.	roundrectGrob()
grid.circle()	Draw circles given locations and radii.	circleGrob()
grid.polygon()	Draw polygons given vertexes.	polygonGrob()
grid.path()	Draw single polygon consisting of multiple paths.	pathGrob()
grid.text()	Draw text given strings, locations and rotations.	textGrob()
grid.raster()	Draw bitmap image.	rasterGrob()
grid.curve()	Draw smooth curve between two end points.	curveGrob()
grid.points()	Draw data symbols given locations.	pointsGrob()

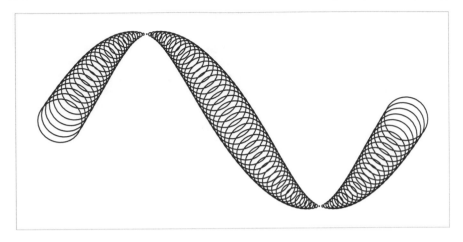

Figure 6.2

Primitive **grid** output. A demonstration of basic graphical output produced using a single call to the **grid.circle()** function. There are 100 circles of varying sizes, each at a different (**x, y**) location.

the curve, including how much the curve deviates from a straight line between the points (**curvature**), whether the curve follows a city-block pattern (**square**), and how smooth the curve is (**ncp**). The following code produces three examples: a city-block curve; a smooth, oblique curve; and a curve that is biased toward the start point and swings wider around the corner (see Figure 6.3).

```
> grid.curve(x1=.1, y1=.25, x2=.3, y2=.75)
> grid.curve(x1=.4, y1=.25, x2=.6, y2=.75,
             square=FALSE, ncp=8, curvature=.5)
> grid.curve(x1=.7, y1=.25, x2=.9, y2=.75,
             square=FALSE, angle=45, shape=-1)
```

The **grid.curve()** function use X-splines to make smooth curves between two end points; the **grid.xspline()** function can be used to produce smooth curves relative to any number of control points. The **grid.bezier()** function draws an approximate cubic Bézier curve; it is only approximate because the curve is actually an X-spline that is parameterised to closely approximate a Bézier curve.

All functions that draw lines have an **arrow** argument, which can be used to add arrowheads to either end of the line. The **arrow()** function is used to create a description of the arrowheads, then this is supplied as the value of the **arrow** argument. The following code demonstrates two possible uses (see

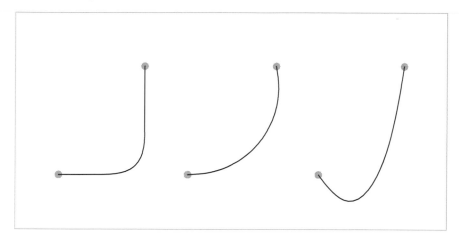

Figure 6.3
Drawing curves between two end points using the `grid.curve()` function: at left is
the default city-block curve; in the middle is a curve that bends less and is symmetric
between the end points; and at right is a curve that is biased toward the starting
point.

Figure 6.4). The call to `grid.lines()` adds an open arrowhead to a single
line and the call to `grid.segments()` adds narrower closed arrowheads to
each of three lines.

```
> angle <- seq(0, 2*pi, length=50)
> grid.lines(x=seq(0.1, 0.5, length=50),
            y=0.5 + 0.3*sin(angle), arrow=arrow())
> grid.segments(6:8/10, 0.2, 7:9/10, 0.8,
                arrow=arrow(angle=15, type="closed"))
```

In simple usage, the `grid.polygon()` function draws a single polygon through
the specified x- and y-locations, automatically joining the last location to
the first to close the polygon. It is possible to produce multiple polygons
from a single call if the `id` argument is specified. In this case, a polygon is
drawn for each set of x- and y-locations corresponding to a different value of
id. The following code demonstrates both usages (see Figure 6.5). The two
`grid.polygon()` calls both use the same x- and y-locations, but the second
call splits the locations into three separate polygons using the `id` argument.

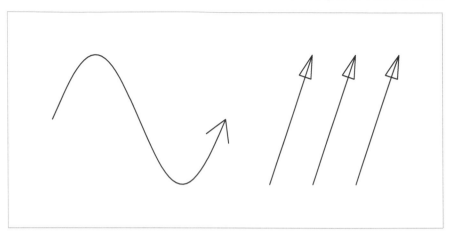

Figure 6.4
Drawing arrows using line-drawing functions. Arrows can be added to the output from `grid.lines()`, `grid.polyline()`, `grid.segments()`, `grid.line.to()`, `grid.xspline()`, and `grid.curve()`. Examples are shown for `grid.lines()` (the sine curve in the left half of the figure) and `grid.segments()` (the three straight lines in the right half of the figure).

```
> angle <- seq(0, 2*pi, length=10)[-10]
> grid.polygon(x=0.25 + 0.15*cos(angle), y=0.5 + 0.3*sin(angle),
              gp=gpar(fill="gray"))
> grid.polygon(x=0.75 + 0.15*cos(angle), y=0.5 + 0.3*sin(angle),
              id=rep(1:3, each=3),
              gp=gpar(fill="gray"))
```

The `grid.path()` function also has an `id` argument, but instead of producing multiple polygons, the result is a single polygon consisting of multiple paths. This can be used to create a shape with an internal hole. The following code shows an example where a polygon shape is created with a rectangular hole in the middle (see Figure 6.6).

```
> angle <- seq(0, 2*pi, length=10)[-10]
> grid.path(x=0.25 + 0.15*cos(angle), y=0.5 + 0.3*sin(angle),
           gp=gpar(fill="gray"))
> grid.path(x=c(0.75 + 0.15*cos(angle), .7, .7, .8, .8),
           y=c(0.5 + 0.3*sin(angle),   .4, .6, .6, .4),
           id=rep(1:2, c(9, 4)),
           gp=gpar(fill="gray"))
```

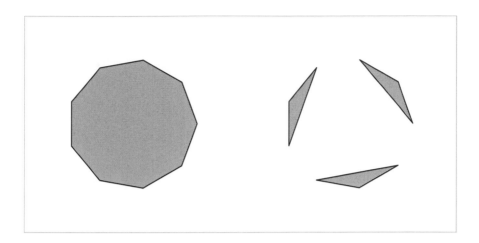

Figure 6.5
Drawing polygons using the `grid.polygon()` function. By default, a single polygon is produced from multiple (x, y) locations (the nonagon on the left), but it is possible to associate subsets of the locations with separate polygons using the `id` argument (the three triangles on the right).

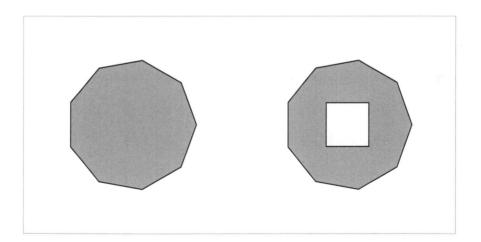

Figure 6.6
Drawing paths using the `grid.path()` function. In simple cases, a single polygon is produced, from multiple (x, y) locations (the nonagon on the left), but it is also possible to associate subsets of the locations with separate subpaths using the `id` argument, which can be used to create holes in the polygon (the shape on the right).

Table 6.2

Graphical utilities in **grid**. These functions draw a small collection of basic shapes or a basic shape that has indirect parameters. As with the graphical primitives, for each function that produces graphical output (left-most column), there is a corresponding function that returns a graphical object containing a description of graphical output instead of producing graphical output (right-most column). The latter set of functions is described further in Chapter 7.

Function to Produce Output	Description	Function to Produce Object
grid.xaxis()	Draw x-axis.	xaxisGrob()
grid.yaxis()	Draw y-axis.	yaxisGrob()
grid.abline()	Draw a line given slope and intercept.	ablineGrob
grid.grill()	Draw vertical and horizontal lines.	grillGrob
grid.function	Draw a curve defined by a function.	functionGrob

The `grid.points()` function draws small shapes as data symbols at the specified (`x, y`) locations. The `pch` argument specifies the data symbol shape as an integer (e.g., `0` means an open square and `1` means an open circle) or as a single character (see Section 10.3).

The `grid.raster()` function draws a bitmap image. The bitmap image can be specified as a vector, matrix, or array. Chapter 11 describes ways to source an image from an external file.

6.2.1   Graphical utilities

In addition to the most basic shapes, **grid** provides some slightly higher level drawing functions. These are listed in Table 6.2 and each one is briefly described in this section.

The `grid.xaxis()` and `grid.yaxis()` functions are not simple graphical primitives because they produce output consisting of both lines and text. The main argument to these functions is the `at` argument. This is used to specify where tick marks should be placed. If the argument is not specified, sensible tick marks are drawn based on the current scales in effect (see Section 6.5 for information about viewport scales). The values specified for the `at` argument are always relative to the current scales (see the concept of the `"native"` coordinate system in Section 6.3). These functions are much less flexible and general than the base `axis()` function. For example, they do not provide automatic support for generating labels from time-based or date-based `at` locations.

The remaining functions all draw one or more lines. The `grid.abline()` function draws a straight line based on intercept and slope values, ensuring that the line starts and ends on an edge of the current viewport. The `grid.grill()` function draws a set of vertical and horizontal lines, based on just a set of horizontal and vertical values, respectively. The `grid.function()` function draws a series of line segments based on a function, which must take a single argument, `x`, and must return a list with `x` and `y` components.

6.2.2 Standard arguments

All primitive graphics functions accept a `gp` argument that allows control over aspects such as the color and line type of the relevant output. For example, the following code specifies that the boundary of the rectangle should be dashed and colored red.

```
> grid.rect(gp=gpar(col="red", lty="dashed"))
```

Section 6.4 provides more information about setting graphical parameters.

All primitive graphics functions also accept a `vp` argument that can be used to specify a viewport in which to draw the relevant output. The following code shows a simple example of the syntax (the result is a rectangle drawn in the left half of the page); Section 6.5 describes viewports and the use of `vp` arguments in full detail.

```
> grid.rect(vp=viewport(x=0, width=0.5, just="left"))
```

Finally, all primitive graphics functions also accept a `name` argument. This can be used to identify the graphical object produced by the function. It is useful for editing graphical output and when working with graphical objects (see Chapter 7). The following code demonstrates how to associate a name with a rectangle.

```
> grid.rect(name="myrect")
```

6.2.3 Clipping

The `grid.clip()` function is not really a graphical primitive because it does not draw anything. Instead, this function specifies a clipping rectangle. After this function has been called, any subsequent drawing will only be visible if it occurs inside the clipping rectangle.

The clipping rectangle can be reset by calling `grid.clip()` again or by changing the drawing viewport (see Section 6.5, especially Section 6.5.2).

6.3   Coordinate systems

When drawing in **grid**, there are always a large number of coordinate systems available for specifying the locations and sizes of graphical output. For example, it is possible to specify an x-location as a proportion of the width of the drawing region, or as a number of inches (or centimeters, or millimeters) from the left-hand edge of the drawing region, or relative to the current x-axis scale. The full set of coordinate systems available is shown in Table 6.3. The meaning of some of these will only become clear with an understanding of viewports (Section 6.5) and graphical objects (Chapter 7).*

With so many coordinate systems available, it is necessary to specify which coordinate system a location or size refers to. This is the purpose of the `unit()` function. This function creates an object of class `"unit"` (hereafter referred to simply as a *unit*), which acts very much like a normal `numeric` object — it is possible to perform basic operations such as subsetting units, and adding and subtracting units.

Each value in a unit can be associated with a different coordinate system and each location and dimension of a graphical object is a separate unit so, for example, a rectangle can have its x-location, y-location, width, and height all specified relative to different coordinate systems.

The following pieces of code demonstrate some of the flexibility of **grid** units. The first code examples show some different uses of the `unit()` function: a single value is associated with a coordinate system, then several values are associated with a coordinate system (notice the recycling of the coordinate system), then several values are associated with different coordinate systems.

```
> unit(1, "mm")
```

```
[1] 1mm
```

---

*Absolute units, such as inches, may not be rendered with full accuracy in all output formats (see the footnote on page 98).

Table 6.3
The full set of coordinate systems available in **grid**.

Coordinate System Name	Description
`"native"`	Locations and sizes are relative to the x- and y-scales for the current viewport.
`"npc"`	Normalized Parent Coordinates. Treats the bottom-left corner of the current viewport as the location $(0,0)$ and the top-right corner as $(1,1)$.
`"snpc"`	Square Normalized Parent Coordinates. Locations and sizes are expressed as a proportion of the *smaller* of the width and height of the current viewport.
`"in"`	Locations and sizes are in terms of physical inches. For locations, $(0,0)$ is at the bottom-left of the viewport.
`"cm"`	Same as `"in"`, except in centimeters.
`"mm"`	Millimeters.
`"pt"`	Points. There are 72.27 points per inch.
`"bigpts"`	Big points. There are 72 big points per inch.
`"picas"`	Picas. There are 12 points per pica.
`"dida"`	Dida. 1157 dida equals 1238 points.
`"cicero"`	Cicero. There are 12 dida per cicero.
`"scaledpts"`	Scaled points. There are 65536 scaled points per point.
`"char"`	Locations and sizes are specified in terms of multiples of the current nominal font size (dependent on the current `fontsize` and `cex`).
`"line"`	Locations and sizes are specified in terms of multiples of the height of a line of text (dependent on the current `fontsize`, `cex`, and `lineheight`).
`"strwidth"` `"strheight"`	Locations and sizes are expressed as multiples of the width (or height) of a given string (dependent on the string and the current `fontsize`, `cex`, `fontfamily`, and `fontface`).
`"grobx"` `"groby"`	Locations and sizes are expressed as multiples of the x- or y-location on the boundary of a given graphical object (dependent on the type, location, and graphical settings of the graphical object).
`"grobwidth"` `"grobheight"`	Locations and sizes are expressed as multiples of the width (or height) of a given graphical object (dependent on the type, location, and graphical settings of the graphical object).

```
> unit(1:4, "mm")
```

[1] 1mm 2mm 3mm 4mm

```
> unit(1:4, c("npc", "mm", "native", "line"))
```

[1] 1npc   2mm   3native 4line

The next code examples show how units can be manipulated in many of the ways that normal numeric vectors can: firstly by subsetting, then simple arithmetic (again notice the recycling), then finally the use of a summary function (`max()` in this case).

```
> unit(1:4, "mm")[2:3]
```

[1] 2mm 3mm

```
> unit(1, "npc") - unit(1:4, "mm")
```

[1] 1npc-1mm 1npc-2mm 1npc-3mm 1npc-4mm

```
> max(unit(1:4, c("npc", "mm", "native", "line")))
```

[1] max(1npc, 2mm, 3native, 4line)

Some operations on units are not as straightforward as with numeric vectors, but require the use of functions written specifically for units. For example, units must be concatenated (in the sense of the `c()` function) using `unit.c()`.

The following code provides an example of using units to locate and size a rectangle. The rectangle is at a location 40% of the way across the drawing region and 1 inch from the bottom of the drawing region. It is as wide as the text "very snug", and it is one line of text high (see Figure 6.7).

```
> grid.rect(x=unit(0.4, "npc"), y=unit(1, "in"),
            width=stringWidth("very snug"),
            height=unit(1, "line"),
            just=c("left", "bottom"))
```

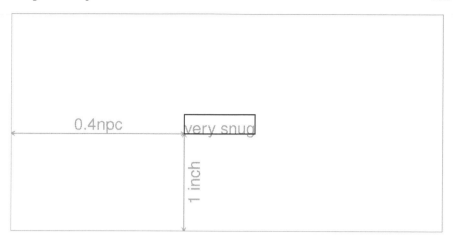

Figure 6.7
A demonstration of **grid** units. A diagram demonstrating how graphical output can be located and sized using **grid** units to associate numeric values with different coordinate systems. The gray border represents the current viewport. A black rectangle has been drawn with its bottom-left corner 40% of the way across the current viewport and 1 inch above the bottom of the current viewport. The rectangle is 1 line of text high and as wide as the text "very snug" (as it would be drawn in the current font).

6.3.1   Conversion functions

As demonstrated in the previous section, a unit is not simply a numeric value. Units only reduce to a simple numeric value (a physical location on a graphics device) when drawing occurs. A consequence of this is that a unit can mean very different things, depending on when it gets drawn (this should become more apparent with an understanding of graphical parameters in Section 6.4 and viewports in Section 6.5).

In some cases, it can be useful to convert a unit to a simple numeric value. For example, it is sometimes necessary to know the current scale limits for numerical calculations. There are several functions that can assist with this problem: `convertX()`, `convertY()`, `convertWidth()`, `convertHeight()`, and `convertUnit()`. The following code shows a calculation of the current page height in inches.

```
> convertHeight(unit(1, "npc"), "in")
```

```
[1] 7in
```

WARNING: These conversion functions must be used with care. The output from these functions is only valid for the current page or screen size. If, for example, a window on screen is resized, or output is copied from the screen to a file format with a different physical size, these calculations may no longer be correct. In other words, only rely on these functions when it is known that the size of the screen will not change. The discussion on the use of these functions in `makeContent()` methods and the functions `grid.record()` and `grid.delay()` is also relevant (see "Calculations during drawing" in Section 8.3.10).

6.3.2   Complex units

A number of coordinate systems in **grid** are *relative* in the sense that a value is interpreted as a multiple of the location or size of some other object. These units include `"strwidth"`, `"strheight"`, `"grobx"`, `"groby"`, `"grobwidth"`, and `"grobheight"` and there are two peculiarities of these sorts of coordinate systems that require further explanation. In the first two cases, `"strwidth"` and `"strheight"` units, the other object is just a text string (e.g., `"a label"`), but in the latter four cases, the other object can be any graphical object (see Chapter 7). It is necessary to specify the other object when generating a unit for these coordinate systems and this is achieved via the `data` argument. The following code shows some simple examples.

```
> unit(1, "strwidth", "some text")
```

```
[1] 1strwidth
```

```
> unit(1, "grobwidth", textGrob("some text"))
```

```
[1] 1grobwidth
```

A more convenient interface for generating units, when all values are relative to a single coordinate system, is also available via the `stringWidth()`, `stringHeight()`, `grobX()`, `grobY()`, `grobWidth()`, and `grobHeight()` functions. The following code is equivalent to the previous example.

```
> stringWidth("some text")
```

```
[1] 1strwidth
```

```
> grobWidth(textGrob("some text"))
```

```
[1] 1grobwidth
```

In this particular example, the `"strwidth"` and `"grobwidth"` units will be identical as they are based on identical pieces of text. The difference is that a graphical object can contain not only the text to draw, but also other information that may affect the size of the text, such as the font family and size.

In the following code, the two units are no longer identical because the `text` grob represents text drawn at font size of 18, whereas the simple string represents text at the default size of 10. The `convertWidth()` function is used to demonstrate the difference.

```
> convertWidth(stringWidth("some text"), "in")
```

```
[1] 0.715666666666667in
```

```
> convertWidth(grobWidth(textGrob("some text",
                                  gp=gpar(fontsize=18))),
               "in")
```

```
[1] 1.0735in
```

For units that contain multiple values, there must be an object specified for every `"strwidth"`, `"strheight"`, `"grobx"`, `"groby"`, `"grobwidth"`, and `"grobheight"` value. Where there is a mixture of coordinate systems within a unit, a value of `NULL` can be supplied for the coordinate systems that do not require data. The following code demonstrates this.

```
> unit(rep(1, 3), "strwidth", list("one", "two", "three"))
```

```
[1] 1strwidth 1strwidth 1strwidth
```

```
> unit(rep(1, 3),
       c("npc", "strwidth", "grobwidth"),
       list(NULL, "two", textGrob("three")))
```

```
[1] 1npc       1strwidth  1grobwidth
```

Again, there is a simpler interface for straightforward situations.

```
> stringWidth(c("one", "two", "three"))
```

[1] 1strwidth 1strwidth 1strwidth

For "grobx", "groby", "grobwidth", and "grobheight" units, it is also possible to specify the name of a graphical object rather than the graphical object itself. This can be useful for establishing a reference to a graphical object, so that when the named graphical object is modified, the unit is updated for the change. The following code demonstrates this idea. First of all, a text grob is drawn with the name "tgrob".

```
> grid.text("some text", name="tgrob")
```

Next, a unit is created that is based on the width of the grob called "tgrob".

```
> theUnit <- grobWidth("tgrob")
```

The convertWidth() function can be used to show the current value of the unit.

```
> convertWidth(theUnit, "in")
```

[1] 0.715666666666667in

The following code modifies the grob named "tgrob" and convertWidth() is used to show that the value of the unit reflects the new width of the text grob.

```
> grid.edit("tgrob", gp=gpar(fontsize=18))
> convertWidth(theUnit, "in")
```

[1] 1.0735in

See Section 7.11 for more examples of calculating the sizes of graphical objects.

6.4   Controlling the appearance of output

All graphical primitives functions (and the viewport() function; see Section 6.5) have a gp argument that can be used to provide a set of graphical parameters to control the appearance of the graphical output. There is a fixed

Table 6.4

The full set of graphical parameters available in **grid**.

Parameter	Description
col	Color of lines, text, rectangle borders, ...
fill	Color for filling rectangles, circles, polygons, ...
alpha	Alpha blending coefficient for transparency
lwd	Line width
lex	Line width expansion multiplier applied to lwd to obtain final line width
lty	Line type
lineend	Line end style (round, butt, square)
linejoin	Line join style (round, miter, bevel)
linemitre	Line miter limit
cex	Character expansion multiplier applied to fontsize to obtain final font size
fontsize	Size of text (in points)
fontface	Font face (bold, italic, ...)
fontfamily	Font family
lineheight	Multiplier applied to final font size to obtain the height of a line

set of graphical parameters (see Table 6.4), all of which can be specified for all types of graphical output.

The value supplied for the **gp** argument must be an object of class **"gpar"**, which is produced using the **gpar()** function. For example, the following code produces a **gpar** object containing graphical parameter settings controlling color and line type.

```
> gpar(col="red", lty="dashed")

$col
[1] "red"

$lty
[1] "dashed"
```

The function **get.gpar()** can be used to obtain current graphical parameter settings. The following code shows how to query the current line type and fill color. When called with no arguments, the function returns a complete list of current settings.

```
> get.gpar(c("lty", "fill"))
```

```
$lty
[1] "solid"
```

```
$fill
[1] "transparent"
```

A `gpar` object represents an *explicit graphical context* — settings for a small number of specific graphical parameters. The example above produces a graphical context that ensures that the color setting is `"red"` and the line-type setting is `"dashed"`. There is also always an *implicit graphical context* consisting of default settings for all graphical parameters. The implicit graphical context is initialized automatically when we call `grid.newpage()` and can be modified by viewports (see Section 6.5.5) or by gTrees (see Section 7.4.1).*

A graphical primitive will be drawn with graphical parameter settings taken from the implicit graphical context, except where there are explicit graphical parameter settings from the graphical primitive's `gp` argument. For graphical primitives, the explicit graphical context is only in effect for the duration of the drawing of the graphical primitive. The following code example demonstrates these rules.

The default initial implicit graphical context includes settings such as `lty="solid"` and `fill="transparent"`. The first rectangle has an explicit setting `fill="black"` so it only uses the implicit setting `lty="solid"`. The second rectangle has no explicit graphical parameter settings so it uses all of the implicit graphical parameter settings, for example, it has a transparent fill. In particular, it is not at all affected by the explicit settings of the first rectangle (see Figure 6.8).

```
> grid.rect(x=0.33, height=0.7, width=0.2,
            gp=gpar(fill="black"))
> grid.rect(x=0.66, height=0.7, width=0.2)
```

---

*The ideas of implicit and explicit graphical contexts are similar to the specification of settings in Cascading Style Sheets and the graphics state in PostScript.

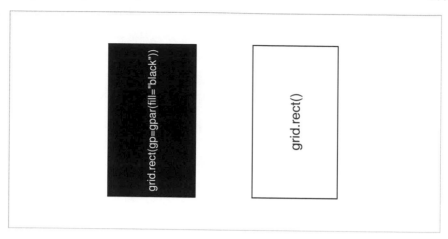

Figure 6.8

Graphical parameters for graphical primitives. The gray rectangle represents the current viewport. The right-hand rectangle has been drawn with no specific graphical parameters so it inherits the defaults for the current viewport (which in this case are a black border and no fill color). The left-hand rectangle has been drawn with a specific fill color of black (it is still drawn with the inherited black border). The graphical parameter settings for one rectangle have no effect on the other rectangle.

6.4.1   Specifying graphical parameter settings

The values that can be specified for colors, line types, line widths, line ends, line joins, and fonts are mostly the same as for the base graphics system. For example, colors can be specified by names such as "red". Chapter 10 describes the specification of graphical parameters in R in complete detail.

One peculiarity to **grid** is that the fontface value can be a name instead of an integer. Table 6.5 shows the possible values.

Many of the parameter names in **grid** are also the same as those in base graphics, though several of the **grid** names are slightly more verbose (e.g., lineend and fontfamily).

In **grid**, the cex value is cumulative. This means that it is multiplied by the previous cex value to obtain a current cex value. The following code shows a simple example. A viewport is pushed with cex=0.5. This means that text will be half size. Next, some text is drawn, also with cex=0.5. This text is drawn quarter size because cex was already 0.5 from the viewport (0.5*0.5 = 0.25).

```
> pushViewport(viewport(gp=gpar(cex=0.5)))
> grid.text("How small do you think?", gp=gpar(cex=0.5))
```

Table 6.5
Possible font face specifications in **grid**.

Integer	Name	Description
1	`"plain"`	Roman or upright face
2	`"bold"`	Bold face
3	`"italic"` or `"oblique"`	Slanted face
4	`"bold.italic"`	Bold and slanted face

The `lex` parameter, which is a multiplier that affects line width, is similarly cumulative.

The `alpha` graphical parameter provides a general alpha-transparency setting. It is a value between 1 (fully opaque) and 0 (fully transparent). The `alpha` value is combined with the alpha channel of colors by multiplying the two and this setting is cumulative like the `cex` setting. The following code shows a simple example. A viewport is pushed with `alpha=0.5`, then a rectangle is drawn using a semitransparent red fill color (alpha channel set to 0.5). The final alpha channel for the fill color is 0.25 (0.5*0.5 = 0.25).

```
> pushViewport(viewport(gp=gpar(alpha=0.5)))
> grid.rect(width=0.5, height=0.5,
            gp=gpar(fill=rgb(1, 0, 0, 0.5)))
```

The **grid** system does not provide any support for fill gradients or patterns, but some effects are possible through judicious use of raster images, graphical primitives, and clipping. Chapter 13 also describes an approach to adding fill patterns to **grid** output.

6.4.2   Vectorized graphical parameter settings

All graphical parameter settings can take a vector of values. Many graphical primitive functions produce multiple primitives as output and graphical parameter settings will be recycled over those primitives. The following code produces 100 circles, cycling through 50 different shades of gray for the circles (see Figure 6.9).

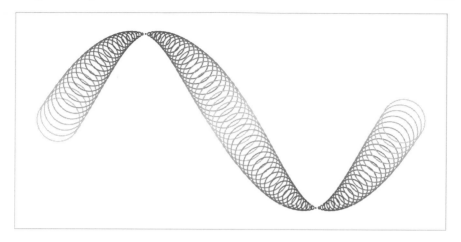

Figure 6.9
Recycling graphical parameters. The 100 circles are drawn by a single function call
with 50 different grays specified for the border color (from a very light gray to a
very dark gray and back to a very light gray). The 50 colors are recycled over the
100 circles so circle *i* gets the same color as circle *i* + 50.

```
> levels <- round(seq(90, 10, length=25))
> grays <- paste("gray", c(levels, rev(levels)), sep="")
> grid.circle(x=seq(0.1, 0.9, length=100),
              y=0.5 + 0.4*sin(seq(0, 2*pi, length=100)),
              r=abs(0.1*cos(seq(0, 2*pi, length=100))),
              gp=gpar(col=grays))
```

The `grid.polygon()` function is a slightly complex case. There are two ways
in which this function will produce multiple polygons: when the `id` argument
is specified *and* when there are `NA` values in the x- or y-locations (see Section 6.6). For `grid.polygon()`, a different graphical parameter will only be
applied to each polygon identified by a different `id`. When a single polygon
(as identified by a single `id` value) is split into multiple subpolygons by `NA`
values, all subpolygons receive the same graphical parameter settings. The
following code demonstrates these rules (see Figure 6.10). The first call to
`grid.polygon()` draws two polygons as specified by the `id` argument. The
`fill` graphical parameter setting contains two colors so the first polygon gets
the first color (gray) and the second polygon gets the second color (white). In
the second call, all that has changed is that an `NA` value has been introduced.
This means that the first polygon as specified by the `id` argument is split into
two separate polygons, but both of these polygons use the same `fill` setting
because they both correspond to an `id` of 1. Both of these polygons get the
first color (gray).

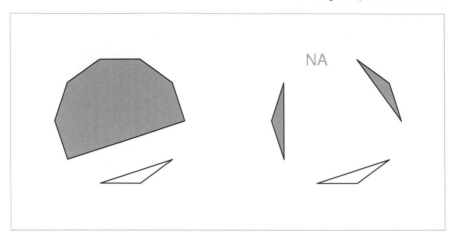

Figure 6.10
Recycling graphical parameters for polygons. On the left, a single function call produces two polygons with different fill colors by specifying an `id` argument and two fill colors. On the right, there are three polygons because an `NA value` has been introduced in the `(x, y)` locations for the polygon, but there are still only two colors specified. The colors are allocated to polygons using the `id` argument and ignoring any `NA` values.

```
> angle <- seq(0, 2*pi, length=11)[-11]
> grid.polygon(x=0.25 + 0.15*cos(angle), y=0.5 + 0.3*sin(angle),
               id=rep(1:2, c(7, 3)),
               gp=gpar(fill=c("gray", "white")))
> angle[4] <- NA
> grid.polygon(x=0.75 + 0.15*cos(angle), y=0.5 + 0.3*sin(angle),
               id=rep(1:2, c(7, 3)),
               gp=gpar(fill=c("gray", "white")))
```

Other functions with an `id` argument, for example, `grid.polyline()` and `grid.xspline()`, obey similar rules. On the other hand, the `grid.path()` function is an exception to the exception because it (conceptually) only ever draws a single shape.

All graphical primitives have a `gp` component, so it is possible to specify any graphical parameter setting for any graphical primitive. This may seem inefficient, and indeed in some cases the values are completely ignored (e.g., text drawing ignores the `lty` setting), but in many cases the values are potentially useful. For example, even when there is no text being drawn, the settings for `fontsize`, `cex`, and `lineheight` are always used to calculate the meaning of `"line"` and `"char"` coordinates. For example, the rectangles produced by the following code are different heights. Both rectangles have their height defined in terms of lines of text so, although no text is drawn, the size of the rectangles can be affected by text-related graphical parameter settings (in this case, the `lineheight` multiplier).

```
> grid.rect(height=unit(1, "lines"))
> grid.rect(height=unit(1, "lines"),
             gp=gpar(lineheight=2))
```

6.5 Viewports

A *viewport* is a rectangular region that provides a context for drawing.

A viewport provides a *drawing context* consisting of both a *geometric context* and a *graphical context*. A geometric context consists of a set of coordinate systems for locating and sizing output and all of the coordinate systems described in Section 6.3 are available within every viewport. A graphical context consists of explicit graphical parameter settings for controlling the appearance of output. This is specified as a `"gpar"` as produced by the `gpar()` function.

By default, **grid** creates a *root* viewport that corresponds to the entire page and, until another viewport is created, drawing occurs within the full extent of the page and using the default graphical parameter settings.*

A new viewport is created using the `viewport()` function. A viewport has a location (given by x and y), a size (given by `width` and `height`), and it is justified relative to its location (according to the value of the `just` argument). The location and size of a viewport are specified in units, so a viewport can be positioned and sized within another viewport in a very flexible manner. The following code creates a viewport that is left-justified at an x-location 0.4 of the way across the drawing region, and bottom-justified 1 centimeter from the bottom of the drawing region. It is as wide as the text `"very very snug indeed"`, and it is six lines of text high. Figure 6.11 shows a diagram representing this viewport.

---

*Warning: some default parameter settings vary between different graphics formats. For example, the default `fill` parameter is usually `"transparent"`, but for PNG output it is `"white"`.

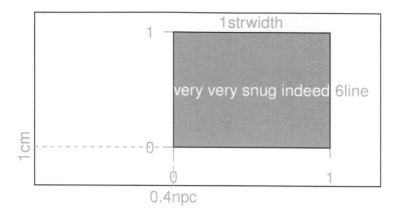

Figure 6.11

A diagram of a simple viewport. A viewport is a rectangular region specified by an (x, y) location, a (width, height) size, and a justification (and possibly a rotation). This diagram shows a viewport that is left-bottom justified 1 centimeter off the bottom of the page and 0.4 of the way across the page. It is six lines of text high and as wide as the text "very very snug indeed."

```
> viewport(x=unit(0.4, "npc"), y=unit(1, "cm"),
           width=stringWidth("very very snug indeed"),
           height=unit(6, "line"),
           just=c("left", "bottom"))
```

viewport[GRID.VP.14]

An important thing to notice in the above example is that the result of the `viewport()` function is an object of class `"viewport"`. No region has actually been created on the page. In order to create regions on the page, a `viewport` object must be *pushed*, as described in the next section.

6.5.1   Pushing, popping, and navigating between viewports

The `pushViewport()` function takes a `viewport` object and uses it to create a region on the graphics device. This region becomes the drawing context for all subsequent graphical output, until the region is removed or another region is defined.

The following code demonstrates this idea (see Figure 6.12). To start with, the entire page and the default graphical parameter settings provide the drawing context. Within this context, the `grid.text()` call draws some text at the top-left corner of the device. A viewport is then pushed, which creates a region 80% as wide as the page, half the height of the page, and rotated at an angle of 10 degrees.* The viewport is given a name, `"vp1"`, which will help us to navigate back to this viewport from another viewport later.

Within the new drawing context defined by the viewport that has been pushed, *exactly the same* `grid.text()` call produces some text at the top-left corner of the viewport. A rectangle is also drawn to make the extent of the new viewport clear.

```
> grid.text("top-left corner", x=unit(1, "mm"),
            y=unit(1, "npc") - unit(1, "mm"),
            just=c("left", "top"))
> pushViewport(viewport(width=0.8, height=0.5, angle=10,
              name="vp1"))
> grid.rect()
> grid.text("top-left corner", x=unit(1, "mm"),
            y=unit(1, "npc") - unit(1, "mm"),
            just=c("left", "top"))
```

The pushing of viewports is entirely general. A viewport is pushed relative to the current drawing context. The following code slightly extends the previous example by pushing a further viewport, exactly like the first, and again drawing text at the top-left corner (see Figure 6.13). The location, size, and rotation of this second viewport are all relative to the context provided by the first viewport. Viewports can be nested like this to any depth.

```
> pushViewport(viewport(width=0.8, height=0.5, angle=10,
              name="vp2"))
> grid.rect()
> grid.text("top-left corner", x=unit(1, "mm"),
            y=unit(1, "npc") - unit(1, "mm"),
            just=c("left", "top"))
```

In **grid**, drawing is always within the context of the current viewport. One way to change the current viewport is to push a viewport (as in the previous examples), but there are other ways too. For a start, it is possible to *pop* a viewport using the `popViewport()` function. This removes the current viewport and the drawing context reverts to whatever it was before the current

---

*It is not often very useful to rotate a viewport, but it helps in this case to dramatize the difference between the drawing regions.

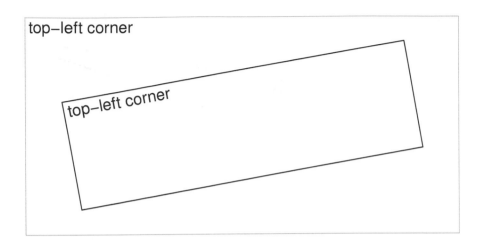

Figure 6.12
Pushing a viewport. Drawing occurs relative to the entire device until a viewport is pushed. For example, some text has been drawn in the top-left corner of the device. Once a viewport has been pushed, output is drawn relative to that viewport. The black rectangle represents a viewport that has been pushed and text has been drawn in the top-left corner of that viewport.

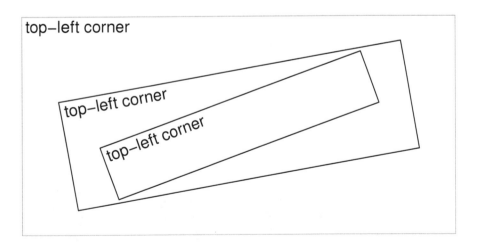

Figure 6.13
Pushing several viewports. Viewports are pushed relative to the current viewport. Here, a second viewport has been pushed relative to the viewport that was pushed in Figure 6.12. Again, text has been drawn in the top-left corner.

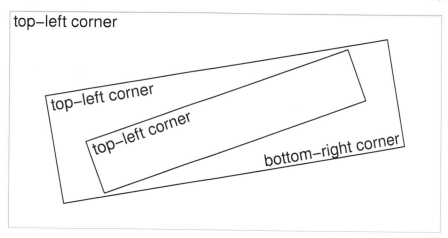

Figure 6.14
Popping a viewport. When a viewport is popped, the drawing context reverts to
the parent viewport. In this figure, the second viewport (pushed in Figure 6.13) has
been popped to go back to the first viewport (pushed in Figure 6.12). This time
text has been drawn in the bottom-right corner.

viewport was pushed. It is illegal to pop the top-most viewport, and trying
to do so will result in an error.

The following code demonstrates popping viewports (see Figure 6.14). The
call to popViewport() removes the last viewport that was created on the page.
Text is drawn at the bottom-right of the resulting drawing region (which has
reverted back to being the first viewport that was pushed).

```
> popViewport()
> grid.text("bottom-right corner",
            x=unit(1, "npc") - unit(1, "mm"),
            y=unit(1, "mm"), just=c("right", "bottom"))
```

The popViewport() function has an integer argument n that specifies how
many viewports to pop. The default is 1, but several viewports can be popped
at once by specifying a larger value. The special value of 0 means that all
viewports should be popped. In other words, the drawing context should
revert to the entire device and the default graphical parameter settings.

Another way to change the current viewport is by using the upViewport()
and downViewport() functions. The upViewport() function is similar to
popViewport() in that the drawing context reverts to whatever it was prior to
the current viewport being pushed. The difference is that upViewport() does
not remove the current viewport from the page. This difference is significant

because it means that a viewport can be revisited without having to push it again. Revisiting a viewport is faster than pushing a viewport and it allows the creation of viewport regions to be separated from the production of output (see "viewport paths" in Section 6.5.3 and Chapter 8).

A viewport can be revisited using the downViewport() function. This function has an argument **name** that can be used to specify the name of an existing viewport. The result of downViewport() is to make the named viewport the current drawing context. The following code demonstrates the use of upViewport() and downViewport() (see Figure 6.15).

A call to upViewport() is made, which reverts the drawing context to the top-level (root) viewport (the entire page) and text is drawn in the bottom-right corner (recall that prior to this navigation the current viewport was the first viewport that was pushed). The downViewport() function is then used to navigate back down to the viewport that was first pushed and a second border is drawn around this viewport. The viewport to navigate down to is specified by its name, "vp1".

```
> upViewport()
> grid.text("bottom-right corner",
            x=unit(1, "npc") - unit(1, "mm"),
            y=unit(1, "mm"), just=c("right", "bottom"))
> downViewport("vp1")
> grid.rect(width=unit(1, "npc") + unit(2, "mm"),
            height=unit(1, "npc") + unit(2, "mm"))
```

There is also a seekViewport() function that can be used to travel across the viewport tree. This can be convenient for interactive use, but the result is less predictable, so it is less suitable for use in writing **grid** functions for others to use. The call seekViewport("avp") is equivalent to upViewport(0); downViewport("avp").

Drawing between viewports

Sometimes it is useful to be able to locate graphical output relative to more than one viewport. One way to do this in **grid** is via the grid.move.to() and grid.line.to() functions. It is possible to call grid.move.to() within one viewport, change viewports, and call grid.line.to().

Another approach is to use the grid.null() function. This is a special graphical primitive that does not draw anything, but it draws nothing at a very specific location. Through the use of the functions grobX() and grobY() this makes it possible to perform drawing relative to one or more invisible loca-

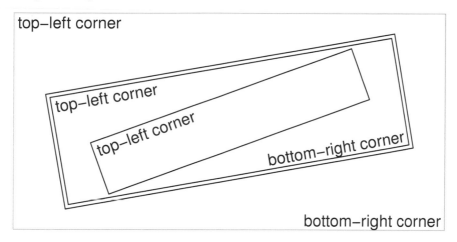

Figure 6.15
Navigating between viewports. Rather than popping a viewport, it is possible to navigate up from a viewport (and leave the viewport on the device). Here navigation has occurred from the first viewport to revert the drawing context to the entire device and text has been drawn in the bottom-right corner. Next, there has been a navigation down to the first viewport again and a second border has been drawn around the outside of the viewport.

tions, represented by one or more "null" grobs, which can be located in one or more different viewports. Section 7.11 has an example of this approach.

6.5.2 Clipping to viewports

Drawing can be restricted to only the interior of the current viewport (*clipped* to the viewport) by specifying the `clip` argument to the `viewport()` function. This argument has three values: `"on"` indicates that output should be clipped to the current viewport; `"off"` indicates that output should not be clipped at all; `"inherit"` means that the clipping region of the previous viewport should be used (this may not have been set by the previous viewport if that viewport's `clip` argument was also `"inherit"`). The following code provides a simple example (see Figure 6.16). A viewport is pushed with clipping on and a circle with a very thick black border is drawn relative to the viewport. A rectangle is also drawn to show the extent of the viewport. The circle partially extends beyond the limits of the viewport, so only those parts of the circle that lie within the viewport are drawn.

```
> pushViewport(viewport(width=.5, height=.5, clip="on"))
> grid.rect()
> grid.circle(r=.7, gp=gpar(lwd=20))
```

Next, another viewport is pushed and this viewport just inherits the clipping region from the first viewport. Another circle is drawn, this time with a gray and slightly thinner border and again the circle is clipped to the viewport.

```
> pushViewport(viewport(clip="inherit"))
> grid.circle(r=.7, gp=gpar(lwd=10, col="gray"))
```

Finally, a third viewport is pushed with clipping turned off. Now, when a third circle is drawn (with a thin, black border) all of the circle is drawn, even though parts of the circle extend beyond the viewport.

```
> pushViewport(viewport(clip="off"))
> grid.circle(r=.7)
> popViewport(3)
```

6.5.3 Viewport lists, stacks, and trees

It can be convenient to work with several viewports at once and there are several facilities for doing this in **grid**. The `pushViewport()` function will accept multiple arguments and will push the specified viewports one after another. For example, the fourth expression below is a shorter equivalent version of the first three expressions.

```
> pushViewport(vp1)
> pushViewport(vp2)
> pushViewport(vp3)
```

```
> pushViewport(vp1, vp2, vp3)
```

The `pushViewport()` function will also accept objects that contain several viewports: viewport lists, viewport stacks, and viewport trees. The function `vpList()` creates a list of viewports and these are pushed "in parallel." The first viewport in the list is pushed, then **grid** navigates back up before the next viewport in the list is pushed. The `vpStack()` function creates a stack of viewports and these are pushed "in series." Pushing a stack of viewports is exactly the same as specifying the viewports as multiple arguments

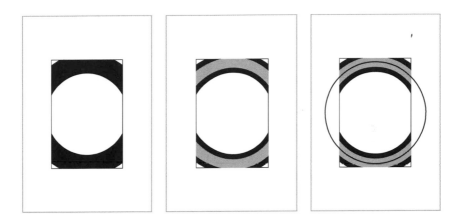

Figure 6.16
Clipping output in viewports. When a viewport is pushed, output can be clipped
to that viewport, or the clipping region can be left in its current state, or clipping
can be turned off entirely. The left panel shows the result of pushing a viewport
(the black rectangle) with clipping on. A circle is drawn with a very thick black
border and it gets clipped. In the middle panel, the left panel is repeated and then
another viewport is pushed (in the same location) with clipping left as it was. A
second circle is drawn with a slightly thinner gray border and it is also clipped. In
the right panel, the middle panel is repeated and then a third viewport is pushed,
which turns clipping off. A circle is drawn with a thin black border and this circle
is not clipped.

to `pushViewport()`. The `vpTree()` function creates a tree of viewports that consists of a parent viewport and any number of child viewports. The parent viewport is pushed first, then the child viewports are pushed in parallel within the parent.

The current set of viewports that have been pushed on the current device constitute a viewport tree and the `current.vpTree()` function prints out a representation of the current viewport tree. The following code demonstrates the output from `current.vpTree()` and the difference between lists, stacks, and trees of viewports. First of all, some (trivial) viewports are created to work with.

```
> vp1 <- viewport(name="A")
> vp2 <- viewport(name="B")
> vp3 <- viewport(name="C")
```

The next piece of code shows these three viewports pushed as a list. The output of `current.vpTree()` shows the root viewport (which represents the entire device) and then all three viewports as children of the root viewport. A graph of the resulting viewport tree is shown in Figure 6.17 (top-left).

```
> pushViewport(vpList(vp1, vp2, vp3))
> current.vpTree()
```

viewport[ROOT]->(viewport[A], viewport[B], viewport[C])

This next code pushes the three viewports as a stack. The viewport `vp1` is now the only child of the root viewport with `vp2` a child of `vp1`, and `vp3` a child of `vp2`. A graph of the resulting viewport tree is shown in Figure 6.17 (top-right).

```
> grid.newpage()
> pushViewport(vpStack(vp1, vp2, vp3))
> current.vpTree()
```

viewport[ROOT]->(viewport[A]->(viewport[B]->(viewport[C])))

Finally, the three viewports are pushed as a tree, with `vp1` as the parent and `vp2` and `vp3` as its children. A graph of the resulting viewport tree is shown in Figure 6.17 (bottom-left).

```
> grid.newpage()
> pushViewport(vpTree(vp1, vpList(vp2, vp3)))
> current.vpTree()
```

```
viewport[ROOT]->(viewport[A]->(viewport[B], viewport[C]))
```

As with single viewports, viewport lists, stacks, and trees can be provided as the **vp** argument for graphical functions (see Section 6.5.4).

Viewport paths

The downViewport() function, by default, searches down the current viewport tree as far as is necessary to find a given viewport name. This is convenient for interactive use, but can be ambiguous if there is more than one viewport with the same name in the viewport tree.

The **grid** system provides the concept of a *viewport path* to resolve such ambiguity. A viewport path is an ordered list of viewport names, which specify a series of parent-child relations. A viewport path is created using the vpPath() function. For example, the following code produces a viewport path that specifies a viewport called "C" with a parent called "B", which in turn has a parent called "A".

```
> vpPath("A", "B", "C")
```

```
A::B::C
```

For convenience in interactive use, a viewport path may be specified directly as a string. For example, the previous viewport path could be specified simply as "C". However, the vpPath() function should be used when writing graphics functions for others to use.

The **name** argument to the downViewport() function will accept a viewport path, in which case it searches for a viewport that matches the entire path. The **strict** argument to downViewport() ensures that a viewport will only be found if the full viewport path is found, *starting from the current location in the viewport tree.*

6.5.4   Viewports as arguments to graphical primitives

As mentioned in Section 6.2.2, a viewport may be specified as an argument to functions that produce graphical output (via an argument called **vp**). When a viewport is specified in this way, the viewport gets pushed before the graphical output is produced and the viewport is popped again afterward. To make this completely clear, the following two code segments are identical. First of all, a simple viewport is defined.

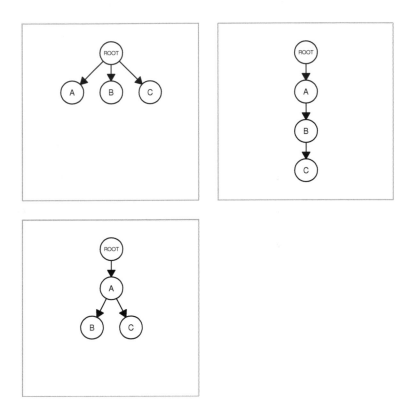

Figure 6.17

Viewport lists, stacks, and trees. There is always a ROOT viewport. At top-left, a *list* of three viewports has been pushed. At top-right, a *stack* of three viewports has been pushed. At bottom-left, a *tree* of three viewports has been pushed (where the tree consists of a parent with two children).

```
> vp1 <- viewport(width=0.5, height=0.5, name="vp1")
```

The next code explicitly pushes the viewport, draws some text, then pops the viewport.

```
> pushViewport(vp1)
> grid.text("Text drawn in a viewport")
> popViewport()
```

This next piece of code does the same thing in a single call.

```
> grid.text("Text drawn in a viewport", vp=vp1)
```

It is also possible to specify the name of a viewport (or a viewport path) for a vp argument. In this case, the name (or path) is used to navigate down to the viewport, via a call to downViewport(), and then back up again afterward, via a call to upViewport(). This promotes the practice of pushing viewports once, then specifying where to draw different output by simply naming the appropriate viewport. The following code does the same thing as the previous example, but leaves the viewport intact (so that it can be used for further drawing).

```
> pushViewport(vp1)
> upViewport()
> grid.text("Text drawn in a viewport", vp="vp1")
```

This feature is also very useful when annotating a plot produced by a high-level graphics function. As long as the graphics function names the viewports that it creates and does not pop them, it is possible to revisit the viewports to add further output. This is what both **lattice** and **ggplot2** do and examples of this sort of annotation are given in Section 6.8. This approach to writing high-level **grid** functions is discussed further in Chapter 8.

6.5.5 Graphical parameter settings in viewports

A viewport can have graphical parameter settings associated with it via the gp argument to viewport(). When a viewport has graphical parameter settings, those settings affect all graphical objects drawn within the viewport, and all other viewports pushed within the viewport, unless the graphical objects or the other viewports specify their own graphical parameter setting. In other words, the graphical parameter settings for a viewport modify the implicit graphical context (see page 198).

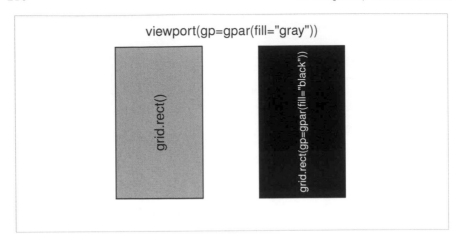

Figure 6.18
The inheritance of viewport graphical parameters. A diagram demonstrating how
viewport graphical parameter settings are inherited by graphical output within the
viewport. The viewport sets the default fill color to gray. The left-hand rectangle
specifies no fill color itself, so it is filled with gray. The right-hand rectangle specifies
a black fill color that overrides the viewport setting.

The following code demonstrates this rule. A viewport is pushed that has
a `fill="gray"` setting. A rectangle with no graphical parameter settings is
drawn within that viewport and this rectangle "inherits" the `fill="gray"`
setting. Next, another rectangle is drawn with its own `fill` setting, so it does
not inherit the viewport setting (see Figure 6.18).

```
> pushViewport(viewport(gp=gpar(fill="gray")))
> grid.rect(x=0.33, height=0.7, width=0.2)
> grid.rect(x=0.66, height=0.7, width=0.2,
            gp=gpar(fill="black"))
> popViewport()
```

The graphical parameter settings in a viewport only affect other viewports and
graphical output within that viewport. The settings do not affect the viewport
itself. For example, parameters controlling the size of text (`fontsize`, `cex`,
etc.) do not affect the meaning of `"line"` units when determining the location
and size of the viewport, but they will affect the location and size of other
viewports or graphical output within the viewport. A layout (see Section
6.5.6) counts as being within the viewport (i.e., it is affected by the graphical
parameter settings of the viewport).

If there are multiple values for a graphical parameter setting, only the first is
used when determining the location and size of a viewport.

6.5.6 Layouts

A viewport can have a *layout* specified via the `layout` argument. A layout in **grid** is similar to the same concept in traditional graphics (see Section 3.3.2). It divides the viewport region into several columns and rows, where each column can have a different width and each row can have a different height. For several reasons, however, layouts are much more flexible in **grid**: there are many more coordinate systems for specifying the widths of columns and the heights of rows (see Section 6.3); viewports can occupy overlapping areas within the layout; and each viewport within the viewport tree can have a layout (layouts can be nested). There is also a `just` argument to justify the layout within a viewport when the layout does not occupy the entire viewport region.

Layouts provide a convenient way to position viewports using the standard set of coordinate systems, and provide an extra coordinate system, `"null"`, which is specific to layouts.

The basic idea is that a viewport can be created with a layout and then subsequent viewports can be positioned relative to that layout. In simple cases, this can be just a convenient way to position viewports in a regular grid, but in more complex cases, layouts are the only way to apportion regions. There are very many ways that layouts can be used in **grid**; the following sections attempt to provide a glimpse of the possibilities by demonstrating a series of example uses.

A **grid** layout is created using the function `grid.layout()` (*not* the base function `layout()`).

A simple layout

The following code produces a simple layout with three columns and three rows, where the central cell (row two, column two) is forced to always be square (using the **respect** argument).

```
> vplay <- grid.layout(3, 3,
                respect=rbind(c(0, 0, 0),
                             c(0, 1, 0),
                             c(0, 0, 0)))
```

The next piece of code uses this layout in a viewport. Any subsequent viewports may make use of the layout, or they can ignore it completely.

```
> pushViewport(viewport(layout=vplay))
```

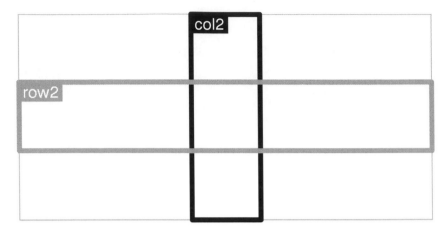

Figure 6.19
Layouts and viewports. Two viewports occupying overlapping regions within a
layout. Each viewport is represented by a rectangle with the viewport name at the
top-left corner. The layout has three columns and three rows with one viewport
occupying all of row two and the other viewport occupying all of column two.

In the next piece of code, two further viewports are pushed within the viewport
with the layout. The `layout.pos.col` and `layout.pos.row` arguments are
used to specify which cells within the layout each viewport should occupy. The
first viewport occupies all of column two and the second viewport occupies all
of row two. This demonstrates that viewports can occupy overlapping regions
within a layout. A rectangle has been drawn within each viewport to show
the region that the viewport occupies (see Figure 6.19).

```
> pushViewport(viewport(layout.pos.col=2, name="col2"))
> upViewport()
> pushViewport(viewport(layout.pos.row=2, name="row2"))
```

A layout with units

This section describes a layout that makes use of **grid** units. In the context
of specifying the widths of columns and the heights of rows for a layout, there
is an additional unit available, the `"null"` unit. All other units (`"cm"`, `"npc"`,
etc.) are allocated first within a layout, then the `"null"` units are used to
divide the remaining space proportionally (see Section 3.3.2). The following
code creates a layout with three columns and three rows. The left column is
one inch wide and the top row is three lines of text high. The remainder of
the current region is divided into two rows of equal height and two columns

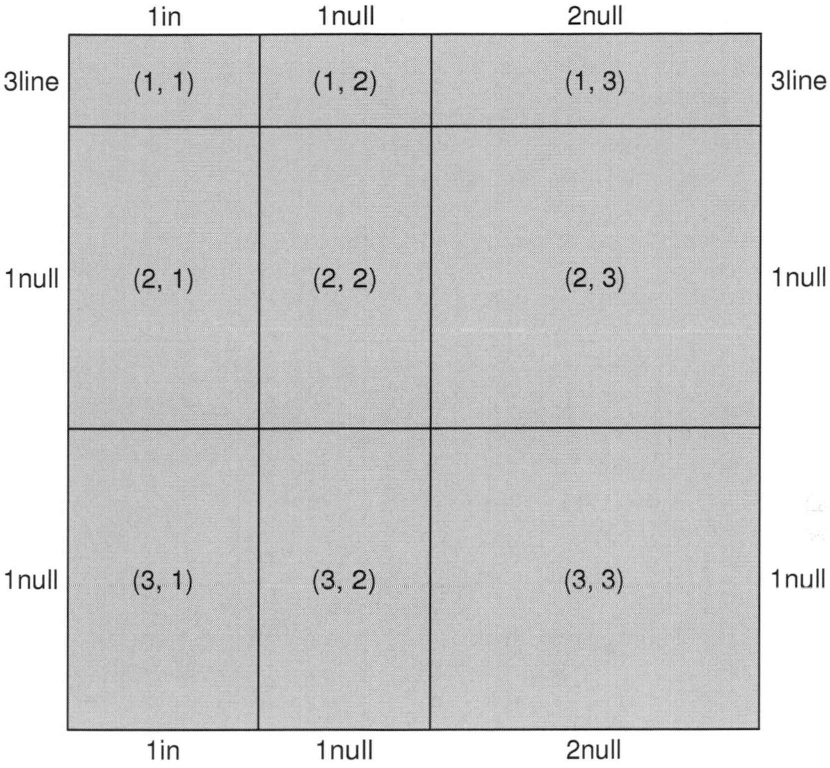

Figure 6.20
Layouts and units. A **grid** layout using a variety of coordinate systems to specify
the widths of columns and the heights of rows.

with the right column twice as wide as the left column (see Figure 6.20).

```
> unitlay <-
    grid.layout(3, 3,
                widths=unit(c(1, 1, 2),
                            c("in", "null", "null")),
                heights=unit(c(3, 1, 1),
                            c("line", "null", "null")))
```

With the use of `"strwidth"` and `"grobwidth"` units it is possible to produce
columns that are just wide enough to fit graphical output that will be drawn
in the column (and similarly for row heights — see Section 7.12).

A nested layout

This section demonstrates the nesting of layouts. The following code defines
a function that includes a trivial use of a layout consisting of two equal-width
columns to produce **grid** output.

```
> gridfun <- function() {
    pushViewport(viewport(layout=grid.layout(1, 2)))
    pushViewport(viewport(layout.pos.col=1))
    grid.rect()
    grid.text("black")
    grid.text("&", x=1)
    popViewport()
    pushViewport(viewport(layout.pos.col=2, clip="on"))
    grid.rect(gp=gpar(fill="black"))
    grid.text("white", gp=gpar(col="white"))
    grid.text("&", x=0, gp=gpar(col="white"))
    popViewport(2)
  }
```

The next piece of code creates a viewport with a layout and places the output
from the above function within a particular cell of that layout (see Figure
6.21).

```
> pushViewport(
    viewport(
      layout=grid.layout(5, 5,
                          widths=unit(c(5, 1, 5, 2, 5),
                                      c("mm", "null", "mm",
                                        "null", "mm")),
                          heights=unit(c(5, 1, 5, 2, 5),
                                       c("mm", "null", "mm",
                                         "null", "mm")))))
> pushViewport(viewport(layout.pos.col=2, layout.pos.row=2))
> gridfun()
> popViewport()
```

The next piece of code calls the function again to draw the same output within
a different cell of the layout.

```
> pushViewport(viewport(layout.pos.col=4, layout.pos.row=4))
> gridfun()
> popViewport(2)
```

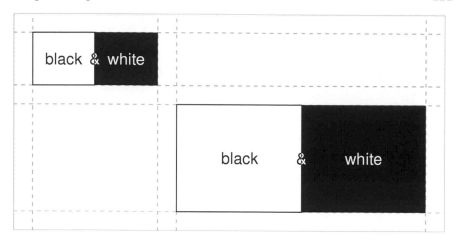

Figure 6.21
Nested layouts. An example of a layout nested within a layout. The black and white squares are drawn within a layout that has two equal-width columns. One instance of the black and white squares has been embedded within cell (2,2) of a layout consisting of five columns and five rows of varying widths and heights (as indicated by the dashed lines). Another instance has been embedded within cell (4,4).

Although the result of this particular example could be achieved using a single layout, what this shows is that it is possible to take **grid** code that makes use of a layout (and may have been written by someone else) and embed it within a layout of your own. A more sophisticated example of this idea, involving **lattice** plots, is given in Section 6.8.2.

6.6 Missing values and non-finite values

Non-finite values are not permitted in the location, size, or scales of a viewport. Viewport scales are checked when a viewport is created, but it is impossible to be certain that locations and sizes are not non-finite when the viewport is created, so this is only checked when the viewport is pushed. Non-finite values result in error messages.

The locations and sizes of graphical objects can be specified as missing values (NA, "NA") or non-finite values (NaN, Inf, -Inf). For most graphical primitives, non-finite values for locations or sizes result in the corresponding primitive not being drawn. For the `grid.line.to()` function, a line segment is only drawn if the previous location and the new location are both not non-finite. For `grid.polygon()`, a non-finite value breaks the polygon into two separate polygons. This break happens within the current polygon as specified by the `id` argument. All polygons with the same `id` receive the same `gp` settings. For line-drawing primitives that are supposed to draw arrowheads, an arrowhead is only drawn if the first or last line segment is drawn.

Figure 6.22 shows the behavior of these primitives where x- and y-locations are seven equally spaced locations around the perimeter of a circle. In the top-left figure, all locations are not non-finite. In each of the other figures, two locations have been made non-finite (indicated in each case by gray text).

Non-finite values for `fontsize`, `lineheight`, and `cex` are silently ignored; the effect is the same as not specifying a parameter setting. This is because there are **grid** units that rely on these parameter settings; ensuring finite values ensures that coordinate system transformations can occur.

6.7   Interactive graphics

The strength of the **grid** system is in the production of static graphics and only very basic support for user interaction is provided via the `grid.locator()` function. This function returns the location of a single mouse click relative to the current viewport. The result is a list containing an `x` and a `y` unit. The `unit` argument can be used to specify the coordinate system that is to be used for the result.

6.8   Customizing lattice plots

The **lattice** package described in Chapter 4 produces complete and very sophisticated plots using **grid**. It makes use of a sometimes large number of viewports to arrange the graphical output. A page of **lattice** output contains a top-level viewport with a quite complex layout that provides space for all of the panels and strips and margins used in the plot. Viewports are created for each panel and for each strip (among other things), and the plot is constructed from a large number of rectangles, lines, text, and data points.

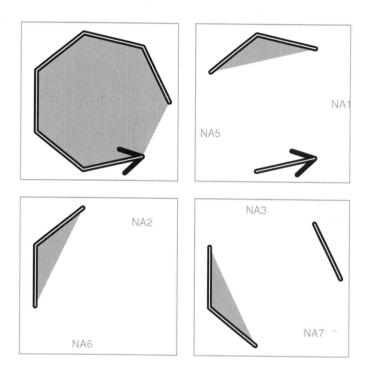

Figure 6.22

Non-finite values for line-to, polygons, and arrows. The effect of non-finite values for grid.line.to(), grid.polygon(), and grid.lines() (with an **arrow** specified). In each panel, a single gray polygon, a single thick black line (with an arrow at the end), and a series of thin white line-tos are drawn through the same set of seven points. In some cases, certain locations have been set to NA (indicated by gray text), which causes the polygon to become cropped, creates gaps in the lines, and can cause the arrowhead to disappear. In the bottom-left panel, the seventh location is not NA, but it produces no output.

In many cases, it is possible to use **lattice** without having to know anything about **grid**. However, a knowledge of **grid** provides a number of more advanced ways to work with **lattice** output (also see Section 7.14).

6.8.1   Adding grid output to lattice output

The functions that **lattice** provides for adding output to panels, for example, `panel.text()` and `panel.points()`, are restricted because they only allow output to be located and sized relative to the `"native"` coordinate system of the panel (i.e., relative to the panel axes). The low-level **grid** graphical primitives provide much more control over the location and size of additional panel output. It is even possible to create and push extra viewports within a panel if desired, although it is very important that they are popped again or **lattice** will get very confused.

In a similar vein, the **grid** functions `upViewport()` and `downViewport()` allow for more flexible navigation of a **lattice** plot compared to the function `trellis.focus()`.

The following code provides an example of `grid.text()` to add output within a **lattice** panel function. This produces a variation on Figure 4.5 with a text label in the top-right corner of each panel to indicate the number of data values in each panel (see Figure 6.23).[*]

```
> xyplot(mpg ~ disp | factor(gear), data=mtcars,
        panel=function(subscripts, ...) {
            grid.text(paste("n =", length(subscripts)),
                      unit(1, "npc") - unit(1, "mm"),
                      unit(1, "npc") - unit(1, "mm"),
                      just=c("right", "top"))
            panel.xyplot(subscripts=subscripts, ...)
        })
```

6.8.2   Adding lattice output to grid output

As well as the advantages of using **grid** functions to add further output to **lattice** plots, an understanding that **lattice** output is really **grid** output makes it possible to embed **lattice** output within **grid** output. The following code provides a simple example where two **lattice** plots are arranged together on a page by drawing them within **grid** viewports (see Figure 6.24).

---

[*]The data are from the `mtcars` data set (see page 130).

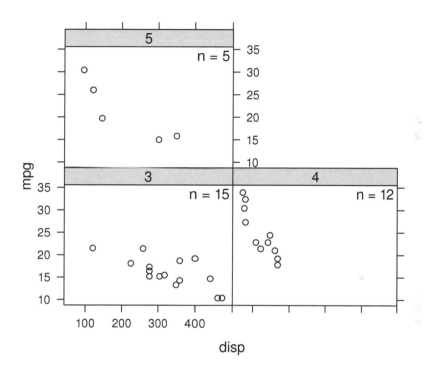

Figure 6.23
Adding **grid** output to a **lattice** plot (the **lattice** plot in Figure 4.5). The **grid**
function `grid.text()` is used within a **lattice** panel function to show the number
of points in each panel.

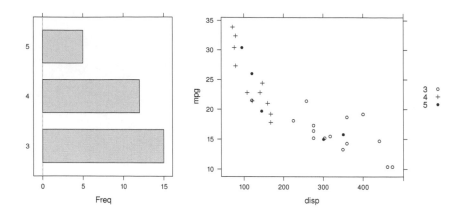

Figure 6.24
Embedding a **lattice** plot within **grid** output. Two **lattice** plots are arranged on a
page by drawing them within **grid** viewports.

```
> grid.newpage()
> pushViewport(viewport(x=0, width=.4, just="left"))
> print(barchart(table(mtcars$gear)),
        newpage=FALSE)
> popViewport()
> pushViewport(viewport(x=.4, width=.6, just="left"))
> print(xyplot(mpg ~ disp, data=mtcars,
               group=gear,
               auto.key=list(space="right")),
        newpage=FALSE)
> popViewport()
```

The viewports are set up using the standard **grid** functions, then the **lattice** plots are drawn within the viewports by explicitly calling `print()` and specifying `newpage=FALSE`.

6.9 Customizing ggplot2 output

Like **lattice**, the **ggplot2** package uses **grid** to do its drawing, which involves
creating a lot of viewports and drawing a lot of graphical primitives. This
means that it is possible to use low-level **grid** functions to manipulate and

add further drawing to **ggplot2** output.

6.9.1 Adding grid output to ggplot2 output

There are two main obstacles to using **grid** functions to add further drawing to **ggplot2** output: we have to call the `grid.force()` function to make the viewports that **ggplot2** produces available (see Section 7.7); and the viewports created by **ggplot2** do not have any knowledge of the x-axis or y-axis scale on the plot, so it is not easy to position extra output relative to the plot scales. A third difficulty is that the viewports that **ggplot2** creates are not named as conveniently as the **lattice** viewports.

Nevertheless, we can still locate further drawing using any of the other **grid** coordinate systems. For example, the following code draws a **ggplot2** scatterplot and then calls `grid.force()` to make the viewports available. The `grid.grep()` function is then used to get the exact name of the "panel" viewport in the plot (see Section 7.2). We then navigate to the panel viewport and place a text label in the top-right corner of the plot (see Figure 6.25).

```
> ggplot(mtcars2, aes(x=disp, y=mpg)) +
      geom_point()


> grid.force()
> panelvp <- grid.grep("panel", grobs=FALSE,
                       viewports=TRUE, grep=TRUE)
> downViewport(panelvp)
> grid.text(paste("n =", nrow(mtcars2)),
            x=unit(1, "npc") - unit(1, "mm"),
            y=unit(1, "npc") - unit(1, "mm"),
            just=c("right", "top"))
```

Chapter 7 provides more information on the ideas and tools required to add **grid** output to **ggplot2** plots.

6.9.2 Adding ggplot2 output to grid output

Similar to **lattice** functions, the **ggplot2** functions create a `"ggplot"` object, which only produces output when it is printed. The printing can be controlled so that, for example, **ggplot2** does not start a new page for the plot. This makes it possible to set up **grid** viewports and draw **ggplot2** output within the viewports.

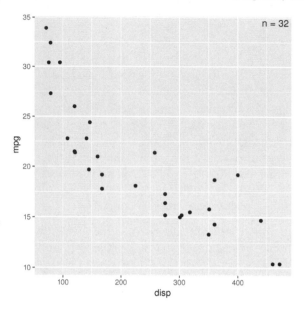

Figure 6.25
Adding **grid** output to a **ggplot2** plot. A text label is added to a **ggplot2** scatter-plot by navigating to the appropriate **ggplot2** viewport and calling grid.text().

The following code demonstrates this idea by drawing a **ggplot2** barplot to the left of a **ggplot2** scatterplot (see Figure 6.26).

```
> grid.newpage()
> pushViewport(viewport(x=0, width=1/3, just="left"))
> print(ggplot(mtcars2, aes(x=trans)) +
        geom_bar(),
        newpage=FALSE)
> popViewport()
> pushViewport(viewport(x=1/3, width=2/3, just="left"))
> print(ggplot(mtcars2, aes(x=disp, y=mpg)) +
        geom_point(aes(color=trans)) +
        scale_color_manual(values=gray(2:1/3)),
        newpage=FALSE)
> popViewport()
```

We can even combine **lattice** plots with **ggplot2** plots, as shown in the fol-lowing code (see Figure 6.27).

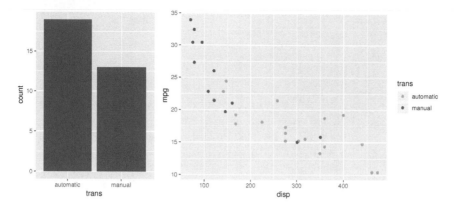

Figure 6.26
Embedding a **ggplot2** plot within **grid** output. Two **ggplot2** plots are drawn
within two **grid** viewports. This is how to get more than one **ggplot2** plot on the
same page.

```
> grid.newpage()
> pushViewport(viewport(x=0, width=.4, just="left"))
> print(ggplot(mtcars2, aes(x=trans)) +
        geom_bar(),
        newpage=FALSE)
> popViewport()
> pushViewport(viewport(x=.4, width=.6, just="left"))
> print(xyplot(mpg ~ disp, data=mtcars,
               group=gear,
               auto.key=list(space="right"),
               par.settings=list(
                   superpose.symbol=list(pch=c(1, 3, 16),
                                         fill="white"))),
        newpage=FALSE)
> popViewport()
```

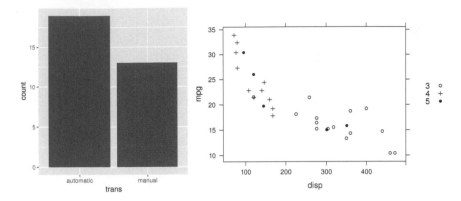

Figure 6.27
Combining a **lattice** plot with a **ggplot2** plot. A **ggplot2** plot is drawn within a **grid** viewport on the left side of the page and a **lattice** plot is drawn within a **grid** viewport on the right side of the page.

Chapter summary

The **grid** package provides a number of functions for producing basic graphical output such as lines, polygons, rectangles, and text, plus some functions for producing slightly more complex output such as data symbols, smooth curves, and axes. Graphical output can be located and sized relative to a large number of coordinate systems and there are a number of graphical parameter settings for controlling the appearance of output, such as colors, fonts, and line types.

Viewports can be created to provide contexts for drawing. A viewport defines a rectangular region on the device and all coordinate systems are available within all viewports. Viewports can be arranged using layouts and nested within one another to produce sophisticated arrangements of graphical output.

Because **lattice** and **ggplot2** output is **grid** output, **grid** functions can be used to add further output to a **ggplot2** or **lattice** plot, and **grid** functions can also be used to control the size and placement of **ggplot2** and **lattice** plots.

7

The grid Graphics Object Model

Chapter preview

This chapter describes how to work with graphical objects (grobs). The main advantage of this approach is that it is possible to modify a scene that was produced using **grid** without having to modify the source code that produced the scene. Because **lattice** and **ggplot2** are built on **grid**, this means it is possible to modify a **ggplot2** or **lattice** plot.

There are also benefits from being able to do such things as ask a piece of graphical output how big it is. For example, this makes it easy to leave space for a legend beside a plot.

Graphical objects can be combined to form larger, hierarchical graphical objects (gTrees). This makes it possible to control the appearance and position of whole groups of graphical objects at once.

This chapter describes the **grid** concepts of grobs and gTrees as well as important functions for accessing, querying, and modifying these objects.

The previous chapter mostly dealt with using **grid** functions to produce graphical output. That knowledge is useful for annotating a plot produced using **grid** (such as a **lattice** plot), for producing one-off or customized plots for your own use, and for writing simple graphics functions.

This chapter addresses **grid** functions for creating and manipulating graphical objects. This information is useful for querying or modifying graphical output that was produced by **grid** (such as a **lattice** plot) and for writing graphical functions and objects for others to use (also see Chapter 8).

7.1   Working with graphical output

This section describes using **grid** to modify graphical output. Every time that
we draw something with **grid**, in addition to producing graphical output, we
create graphical objects, called *grobs*, and **grid** keeps a record of those objects
(called a *display list*). For example, the following code draws a circle (see the
left panel of Figure 7.1).

```
> grid.circle(r=.4, name="mycircle")
```

In addition to producing a circle that we can see, this code generates a
"circle" grob on the display list. We can call the `grid.ls()` function to
see all grobs on the display list.

```
> grid.ls()
```

mycircle

We can also call functions to modify grobs on the display list. This is where
it is important that we gave the grob a name because we use the grob names
to identify which grob we want to modify.

For example, the `grid.edit()` function can be used to modify a grob on the
display list. In the following code, we modify the circle object to change its
fill color (see the middle panel of Figure 7.1). In this case, the `gp` component
of the `circle` grob is being modified. Typically, most arguments that can be
specified when first drawing output can also be used when editing output.

```
> grid.edit("mycircle",
          gp=gpar(fill="grey"))
```

The `grid.remove()` function can be used to remove a grob from the display
list. The following code deletes the output by removing the `circle` object
from the display list (see the right panel of Figure 7.1).

```
> grid.remove("mycircle")
```

Any output produced by **grid** functions can be interacted with in this way,
including output from **lattice** and **ggplot2** functions (see Sections 7.14 and
7.15).

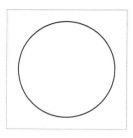  

Figure 7.1
Modifying a `circle` grob. The left panel shows the output produced by a call to
`grid.circle()`, the middle panel shows the result of using `grid.edit()` to modify
the colors of the circles, and the right panel shows the result of using `grid.remove()`
to delete the circles.

Table 7.1
Functions for working with grobs. Functions of the form `grid.*()` access and de-
structively modify grobs on the **grid** display list and affect graphical output. Functions of the form `*Grob()` work with user-level grobs and return grobs as their values
(they have no effect on graphical output).

Function to Work with Output	Description	Function to Work with grobs
`grid.get()`	Returns a copy of one or more grobs	`getGrob()`
`grid.edit()`	Modifies one or more grobs	`editGrob()`
`grid.add()`	Adds a grob to one or more grobs	`addGrob()`
`grid.remove()`	Removes one or more grobs	`removeGrob()`
`grid.set()`	Replaces one or more grobs	`setGrob()`

In addition to editing and removing grobs, we can add grobs and replace grobs;
Table 7.1 shows the main functions for working with grobs on the display list.

We can also disable the **grid** display list, using the `grid.display.list()`
function, in which case no grobs are stored, so these sorts of manipulations
are no longer possible.

7.2   Listing graphical objects

All of the functions that modify grobs on the **grid** display list require the name
of a grob as their first argument. For complex **grid** scenes, such as **lattice**
or **ggplot2** plots, there may be many grobs in the scene and the names of
the grobs may not be known. In these cases, the `grid.ls()` function can be
useful to list the names of the grobs in the current scene.

As an example, the following code draws a simple **lattice** scatterplot (see
Figure 7.2.

```
> xyplot(mpg ~ disp, mtcars)
```

The grobs in this plot are listed with a call to `grid.ls()`. This allows us to
see all of the names of the grobs that **lattice** created.

```
> grid.ls()
```

```
plot_01.background
plot_01.xlab
plot_01.ylab
plot_01.ticks.top.panel.1.1
plot_01.ticks.left.panel.1.1
plot_01.ticklabels.left.panel.1.1
plot_01.ticks.bottom.panel.1.1
plot_01.ticklabels.bottom.panel.1.1
plot_01.ticks.right.panel.1.1
plot_01.xyplot.points.panel.1.1
plot_01.border.panel.1.1
```

We can also list the viewports that **lattice** created and we can show longer
names so that it is easier to distinguish between grobs and viewports.

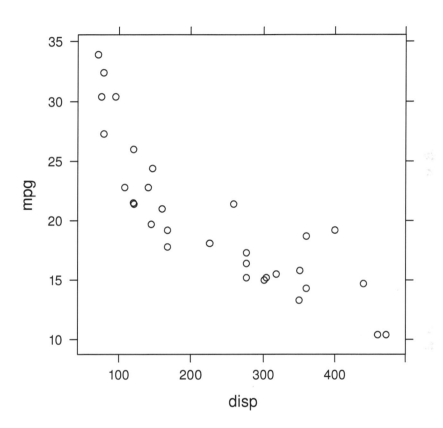

Figure 7.2
Listing grobs using `grid.ls()`. This plot is produced by **lattice**, but it generates a number of **grid** grobs and viewports.

```
> grid.ls(viewports=TRUE, fullNames=TRUE)
```

```
viewport[ROOT]
  rect[plot_01.background]
  viewport[plot_01.toplevel.vp]
    viewport[plot_01.xlab.vp]
      text[plot_01.xlab]
      upViewport[1]
    viewport[plot_01.ylab.vp]
      text[plot_01.ylab]
      upViewport[1]
    viewport[plot_01.figure.vp]
      upViewport[1]
    viewport[plot_01.panel.1.1.vp]
      upViewport[1]
    viewport[plot_01.strip.1.1.off.vp]
      segments[plot_01.ticks.top.panel.1.1]
      upViewport[1]
    viewport[plot_01.strip.left.1.1.off.vp]
      segments[plot_01.ticks.left.panel.1.1]
      text[plot_01.ticklabels.left.panel.1.1]
      upViewport[1]
    viewport[plot_01.panel.1.1.off.vp]
      segments[plot_01.ticks.bottom.panel.1.1]
      text[plot_01.ticklabels.bottom.panel.1.1]
      segments[plot_01.ticks.right.panel.1.1]
      upViewport[1]
    downViewport[plot_01.panel.1.1.vp]
      points[plot_01.xyplot.points.panel.1.1]
      upViewport[1]
    downViewport[plot_01.panel.1.1.off.vp]
      rect[plot_01.border.panel.1.1]
      upViewport[1]
    viewport[plot_01.]
      upViewport[1]
    upViewport[1]
```

In the case of **lattice** plots, all grob and viewport names should be unique and meaningful, but that is not guaranteed to be the case for other code. For example, some of the grobs and viewports in **ggplot2** plots are less obvious (the case of **ggplot2** plots is discussed in more detail in Section 7.15).

When we have a lot of grobs and viewports and the naming system is less clear, the `grid.grep()` function can be used to search for grob or viewport names. For example, using the **lattice** plot above, the following code finds the names of all grobs that contain the word "lab".

```
> grid.grep("lab", grep=TRUE, global=TRUE)
```

```
[[1]]
plot_01.xlab

[[2]]
plot_01.ylab

[[3]]
plot_01.ticklabels.left.panel.1.1

[[4]]
plot_01.ticklabels.bottom.panel.1.1
```

7.3 Selecting graphical objects

All of the functions that modify grobs on the **grid** display list, such as `grid.edit()`, require the name of a grob as their first argument. This selects which grob to modify.

In the simplest case, we want to modify a single grob, but this section shows how to select more than one grob to modify.

To help demonstrate these situations, the following code draws eight concentric `circle` grobs. The first, third, fifth, and seventh circles are named "circle.odd" and the second, fourth, sixth, and eighth circles are named "circle.even". The circles are initially drawn with decreasing shades of gray (see the left panel of Figure 7.3).

```
> suffix <- rep(c("odd", "even"), 4)
> names <- paste0("circle.", suffix)
> names
```

```
[1] "circle.odd"  "circle.even" "circle.odd"  "circle.even"
[5] "circle.odd"  "circle.even" "circle.odd"  "circle.even"
```

```
> for (i in 1:8)
    grid.circle(name=names[i], r=(9 - i)/20,
                gp=gpar(col=NA, fill=gray(i/10)))
```

The function `grid.ls()` shows that we have eight grobs in the current scene.

```
> grid.ls()
```

```
circle.odd
circle.even
circle.odd
circle.even
circle.odd
circle.even
circle.odd
circle.even
```

All of the functions for working with graphical output have a **grep** argument. If we set this to TRUE, then the grob name that we provide as the first argument is treated as a regular expression. There is also a **global** argument and if we set that to TRUE, then *all* matching grobs on the display list (not just the first) will be selected.

Working with the concentric circles that we drew above, the following call to `grid.edit()` makes use of the **global** argument to modify all grobs named `"circle.odd"` and change their fill color to a very dark gray (see the middle panel of Figure 7.3).

```
> grid.edit("circle.odd", gp=gpar(fill="gray10"),
            global=TRUE)
```

A second call to `grid.edit()`, below, makes use of both the **grep** argument and the **global** argument to modify all grobs with names matching the pattern `"circle"` (all of the circles) and change their fill color to a light gray and their border color to a darker gray (see the right panel of Figure 7.3).

```
> grid.edit("circle", gp=gpar(col="gray", fill="gray90"),
            grep=TRUE, global=TRUE)
```

Section 7.15 provides an example of using these more complex selections when modifying a **ggplot2** plot.

There are convenience functions `grid.gget()`, `grid.gedit()`, and `grid.gremove()` that have the **grep** and **global** arguments set to TRUE by default.

Figure 7.3

Editing grobs using `grep` and `global` in `grid.edit()`. The left-hand panel shows eight separate concentric circles, with names alternating between `"circle.odd"` and `"circle.even"`, filled with progressively lighter shades of gray. The middle panel shows the use of the `global` argument to change the fill for all circles named `"circle.odd"` to black. The right-hand panel shows the use of the `grep` and `global` arguments to change all circles whose names match the pattern `"circle"` (all of the circles) to have a light gray fill and a gray border.

7.4 Grob lists, trees, and paths

As well as basic grobs, it is possible to work with a list of grobs (a gList) or several grobs combined together in a tree-like structure (a gTree). A gList is just a list of several grobs (produced by the function `gList()`). A gTree is a grob that can contain other grobs. Two examples of gTree objects are the `xaxis` and `yaxis` objects that are produced by the `grid.xaxis()` and `grid.yaxis()` functions. More complex examples are the gTrees that are produced to draw **ggplot2** plots (see Section 7.15). Another source of gTrees is the `grid.grab()` function (see Section 7.10). This section looks at how to work with gTrees.

When we call the `grid.xaxis()` function, in addition to drawing an axis, we create an `xaxis` grob. This grob contains a high-level description of an axis, plus several child grobs representing the lines and text that make up the axis (see Figure 7.4).

The following code draws an x-axis and creates an `xaxis` grob on the display list (see the left panel of Figure 7.5). The `grid.ls()` function shows that the `axis1` grob has three child grobs. Indenting is used to show that the `major`, `ticks`, and `labels` grobs are children of the `axis1` grob.

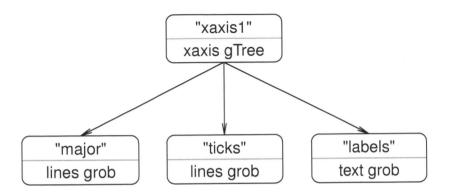

Figure 7.4
The structure of a gTree. A diagram of the structure of an `xaxis` gTree. There is
the `xaxis` gTree itself (here given the name `"xaxis1"`) and there are its children: a
`lines` grob named `"major"`, another `lines` grob named `"ticks"`, and a `text` grob
named `"labels"`.

```
> grid.xaxis(name="axis1", at=1:4/5)
> grid.ls()

axis1
  major
  ticks
  labels
```

The hierarchical structure of gTrees makes it possible to interact with both a
high-level description, as provided by the `xaxis` grob, and a low-level description, as provided by the children of the gTree. The following code demonstrates
an interaction with the high-level description of an `xaxis` grob. The `xaxis`
gTree contains components describing where to put tick marks on the axis and
whether to draw labels and so on. The code below shows the `at` component of
an `xaxis` grob being modified. The `xaxis` grob is designed so that it modifies
its children to match the new high-level description so that only three ticks
are now drawn (see the middle panel of Figure 7.5).

```
> grid.edit("axis1", at=1:3/4)
```

It is also possible to access the children of a gTree. In the case of an `xaxis`,
there are three children: a `lines` grob with the name `"major"`; another `lines`

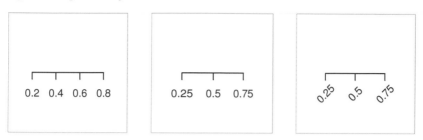

Figure 7.5
Editing a gTree. The left-hand panel shows a basic x-axis, the middle panel shows
the effect of editing the `at` component of the x-axis (all of the tick marks and
labels have changed), and the right-hand panel shows the effect of editing the `rot`
component of the `"labels"` child of the x-axis (only the angle of rotation of the
labels has changed).

grob with the name `"ticks"`; and a `text` grob with the name `"labels"`. Any
of these children can be accessed by specifying the name of the `xaxis` grob
and the name of the child in a *grob path* (gPath). A gPath is like a viewport
path (see Section 6.5.3) — it is just a concatenation of several grob names.
The following code shows how to access the `"labels"` child of the `xaxis` grob
using the `gPath()` function to specify a gPath. The gPath specifies the child
called `"labels"` in the gTree called `"axis1"`. The labels are rotated to 45
degrees (see the right panel of Figure 7.5).

```
> grid.edit(gPath("axis1", "labels"), rot=45)
```

It is also possible to specify a gPath directly as a string, for example `"labels"`,
but this is only recommended for interactive use.

7.4.1 Graphical parameter settings in gTrees

A gTree can have graphical parameter settings associated with it, in which
case, these settings affect all graphical objects that are children of the gTree,
unless the children specify their own graphical parameter setting. In other
words, the graphical parameter settings for a gTree modify the implicit graphical context for the children of the gTree (see page 198).

The following code demonstrates this rule. First, we create an `xaxis` grob,
then we edit the graphical parameter settings of the high-level `"axis2"` gTree
and specify the drawing color to be `"gray"`. This means that all of the children
of the `xaxis`, the lines and labels, will be drawn gray. Finally, we edit the
graphical parameter setting of the low-level tick `"labels"` so that only those
are drawn black (see Figure 7.6).

Figure 7.6
Graphical parameters in a gTree. The left-hand panel shows a basic x-axis, the middle panel shows the effect of editing the **gp** component of the x-axis (all of the tick marks and labels have changed color), and the right-hand panel shows the effect of editing the **gp** component of the **"labels"** child of the x-axis (only the labels have changed color).

```
> grid.xaxis(name="axis2", at=1:4/5)
```

```
> grid.edit("axis2", gp=gpar(col="gray"))
```

```
> grid.edit("labels", gp=gpar(col="black"))
```

Another example of this behavior is given in Section 7.8.

7.5   Searching for grobs

This section provides details about how grob names and gPaths are used to find a grob.

Grobs are stored on the **grid** display list in the order that they are drawn. When searching for a matching name, the functions in Table 7.1 search the display list from the beginning. This means that if there are several grobs whose names are matched, they will be found in the order that they were drawn.

Furthermore, the functions perform a depth-first search. This means that if there is a gTree on the display list, and its name is not matched, then its children are searched for a match before any other grobs on the display list are searched.

The name to search for can be given as a gPath, which makes it possible to explicitly specify a particular child grob of a particular gTree. For example, `"labels"` specifies a grob called `"labels"` that must have a parent called `"axis1"`.

The argument `strict` controls whether a complete match must be found. By default, the `strict` argument is `FALSE`, so in the previous example, the `"labels"` child of `"axis2"` could have been accessed with the expression `grid.get("labels")`. On the other hand, if `strict` is set to `TRUE`, then simply specifying `"labels"` results in no match because there is no top-level grob with the name `"labels"`, as shown by the following code.

```
> grid.edit("labels", strict=TRUE, rot=45)
```

```
Error in
   editDLfromGPath(gPath, specs, strict, grep, global, redraw) :

   'gPath' (labels) not found
```

7.6   Editing graphical context

When a grob is edited using `grid.edit()` or `editGrob()`, the modification of a `gp` component is treated as a special case. Only the graphical parameters that are explicitly given new settings are modified. All other settings remain untouched. The following code provides a simple example.

A circle is drawn with a gray fill color (see the left panel of Figure 7.7), then the border of the circle is made thick (see the middle panel of Figure 7.7) and the fill color remains the same. Finally, the border is changed to a dashed line type, but it stays thick (and the fill remains gray — see the right panel of Figure 7.7).

```
> grid.circle(r=0.3, gp=gpar(fill="gray80"),
              name="mycircle")
> grid.edit("mycircle", gp=gpar(lwd=5))
> grid.edit("mycircle", gp=gpar(lty="dashed"))
```

 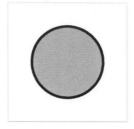 

Figure 7.7
Editing the graphical context. The left-hand panel shows a circle with a solid, thin black border and a gray fill. The middle panel shows the effect of making the border thicker. The important point is that the other features of the circle are not affected (the border is still solid and the fill is still gray). The right-hand panel shows another demonstration of the same idea, with the border now drawn dashed (but the border is still thick and the fill is still gray).

7.7   Forcing graphical objects

It is possible to create a gTree that does not immediately draw its children (see Section 8.3.4) and only decides what to draw when the gTree as a whole is drawn. For example, if we call `grid.xaxis()` without specifying the `at` argument, an `xaxis` gTree is created without any children. This is because the `xaxis` will decide what tick marks to draw only when it is drawn (when it can ask for the `"native"` scale of the viewport that it is drawn within).

The following code provides an example. We first push a viewport and specify an x-axis scale of 0 to 100. A rectangle is drawn to show the location of the viewport and an x-axis is drawn along the bottom edge of the viewport (see Figure 7.8).

```
> pushViewport(viewport(xscale=c(0, 100)))
> grid.rect(name="rect")
> grid.xaxis(name="axis3")
```

If we list the grobs on the display list, there are two, the gray rectangle and the `xaxis`, but the `xaxis` has no children (on the display list).

```
> grid.ls()

rect
axis3
```

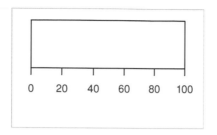

Figure 7.8
An `xaxis` grob drawn within a viewport (represented by the black rectangle) with an x-axis scale from 0 to 100. This axis has an `at` value of `NULL` so it determines its tick mark locations and labels as it is drawn (rather than when it was created).

This means that we cannot directly access or modify the children of the `xaxis`. However, the `grid.force()` function can be used to force the `xaxis` to create its children on the display list, as shown in the following code. The children of the `xaxis` are now visible on the display list, so we can access and modify them.

```
> grid.force()
> grid.ls()

rect
axis3
  major
  ticks
  labels
```

Another, more complex, example of this situation occurs when we create a **ggplot2** plot. The **ggplot2** functions create a gTree with very few children; most grobs are only created as the **ggplot2** plot is drawn and are not recorded on the display list. This is why it is necessary to call `grid.force()` before we can modify the grobs and viewports for a **ggplot2** plot (see Section 7.15).

7.8 Working with graphical objects off-screen

Chapter 6 described **grid** functions that draw graphical output. All of those functions also create grobs representing the drawing and those grobs are stored on the **grid** display list.

We can also create a grob without producing any output. This section describes how to use **grid** to just produce graphical objects (without drawing them). There are functions to create grobs, functions to combine them and to modify them, and the `grid.draw()` function to draw them.

For each **grid** function that produces graphical output, there is a counterpart that produces a graphical object and no graphical output. For example, the counterpart to `grid.circle()` is the function `circleGrob()` (see Table 6.1). Similarly, for each function that works with grobs on the **grid** display list, there is a counterpart for working with grobs off-screen. For example, the counterpart to `grid.edit()` is `editGrob()` (see Table 7.1).

The following example demonstrates the process of creating a grob and working with the grob without drawing it. The code below draws a rectangle that is as wide as a text grob, but the text is not drawn. The function `textGrob()` produces a **text** grob, but does not draw it.

```
> grid.rect(width=grobWidth(textGrob("Some text")))
```

We can also create a grob and modify it before producing any graphical output (i.e., only draw the final result). The following code shows an example involving an **xaxis**. The first expression creates an **xaxis**, but does not draw anything.

```
> ag <- xaxisGrob(at=1:4/5, name="axis4")
```

Because we specified the **at** argument, the **xaxis** children have also been created. The next code shows that the `grid.ls()` function can be given a gTree as its first argument, in which case it will list the children of the gTree.

```
> grid.ls(ag)

axis4
  major
  ticks
  labels
```

Next, we modify the font face for the `"labels"` child of the **xaxis** to be italic. The result is the modified **xaxis** object; still nothing has been drawn to this point.

```
> ag <- editGrob(ag, "labels", gp=gpar(fontface="italic"))
```

Finally, we call the `grid.draw()` function to draw the (modified) `xaxis`.

```
> grid.draw(ag)
```

7.9 Reordering graphical objects

Another way of modifying the children of a gTree is to change the order in which they are drawn, using the `grid.reorder()` function.

In order to demonstrate this tool, the following code creates a gTree with a set of rectangles as its children, one wide and short, one thin and tall, and one square (in that order), and draws it (see the left panel of Figure 7.9).

```
> r1 <- rectGrob(height=.2, gp=gpar(fill="black"), name="r1")
> r2 <- rectGrob(width=.2, gp=gpar(fill="grey"), name="r2")
> r3 <- rectGrob(width=.4, height=.4, gp=gpar(fill="white"),
                 name="r3")
> gt <- gTree(children=gList(r1, r2, r3), name="gt")
> grid.draw(gt)
```

The following call to `grid.reorder()` reverses the order of the children so that the square is drawn first (at the back see the right panel of Figure 7.9).

```
> grid.reorder("gt", c("r3", "r2", "r1"))
```

7.10 Capturing output

In the example above, several grobs were created off-screen and then grouped together as a gTree, which allowed the collection of grobs to be dealt with as a single object.

It is also possible first to *draw* several grobs and *then* to group them. The `grid.grab()` function does this by generating a gTree from all of the grobs in the current page of output. This means that output can be captured even

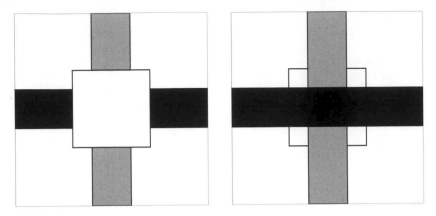

Figure 7.9
Reordering children in a gTree.

from a function that produces very complex output (lots of grobs), such as
a **lattice** plot. For example, the following code draws a **lattice** plot, then
creates a gTree containing all of the grobs in the plot.

```
> bwplot(voice.part ~ height, data=singer)
> bwplotTree <- grid.grab()
```

The `grid.grab()` function actually captures all of the viewports in the current
scene as well as the grobs, so drawing the gTree, as in the following code,
produces exactly the same output as the original plot.

```
> grid.newpage()
> grid.draw(bwplotTree)
```

Another function, `grid.grabExpr()` allows **grid** output to be captured off-
screen. This function takes an R expression and evaluates it. Any drawing
that occurs as a result of evaluating the expression does not produce any
output, but the grobs that would be produced are captured anyway.

The following code provides a simple demonstration. Here a **lattice** plot is
captured without drawing any output.*

```
> grid.grabExpr(print(bwplot(voice.part ~ height, data=singer)))
```

---

*The expression must explicitly `print()` the **lattice** plot because otherwise nothing
would be drawn (see Section 4.1).

```
gTree[GRID.gTree.75]
```

Both the `grid.grab()` and `grid.grabExpr()` functions attempt to create a gTree in a sophisticated way so that it is easier to work with the resulting gTree. Unfortunately, this will not always produce a gTree that will exactly replicate the original output. These functions issue warnings if they detect a situation where output may not be reproduced correctly, and there is a `wrap` argument that can be used to force the functions to produce a gTree that is less sophisticated, but is guaranteed to replicate the original output.

7.11   Querying grobs

Another benefit of having graphical objects (as well as producing graphical output) is that we can query graphical objects to find information about them. Section 6.3.2 already described one way to access this facility with the `grobWidth()` and `grobHeight()` functions, which allow us to calculate the width and height of graphical output. The following code provides a simple example by drawing some text and then a rectangle that is the same width as the text.

```
> grid.text("text", name="t")
> grid.rect(width=grobWidth("t"))
```

A small extension to this idea is that we do not need to draw the text in order to calculate its width. The following code again draws a rectangle that is as wide as a piece of text, but it does not draw the text. Instead it just creates a text grob and queries that.

```
> t <- textGrob("text")
> grid.rect(width=grobWidth(t))
```

This section describes some important extra details about the calculation of grob sizes and the editing of graphical contexts.

7.11.1   Calculating the sizes of grobs

The `"grobwidth"` and `"grobheight"` units, and the `grobWidth()` and `grobHeight()` functions, provide a way to determine the size of a grob.

The most important point about this calculation is that the size of a grob is always calculated relative to the current geometric and graphical context. The following code demonstrates this point. First of all, a `text` grob and a `rect` grob are created, and the dimensions of the `rect` grob are based on the dimensions of the text.[*]

```
> tg1 <- textGrob("Sample")
> rg1 <- rectGrob(x=rep(0.5, 2),
                  width=1.1*grobWidth(tg1),
                  height=1.3*grobHeight(tg1),
                  gp=gpar(col=c("gray60", "white"),
                          lwd=c(3, 1)))
```

Next, these two grobs are drawn in three different settings. In the first setting, the rectangle and the text are drawn in the default geometric and graphical context and the rectangle bounds the text (see the left panel of Figure 7.10).

```
> grid.draw(tg1)
> grid.draw(rg1)
```

In the second setting, the grobs are both drawn within a viewport that has `cex=2`. Both the text and the rectangle are drawn bigger (the calculation of the `"grobwidth"` and `"grobheight"` units takes place in the same context as the drawing of the `text` grob; see the middle panel of Figure 7.10).

```
> pushViewport(viewport(gp=gpar(cex=2)))
> grid.draw(tg1)
> grid.draw(rg1)
> popViewport()
```

In the third setting, the `text` grob is drawn in a different context than the rectangle, so the rectangle's size is "wrong" (see the right panel of Figure 7.10).

```
> pushViewport(viewport(gp=gpar(cex=2)))
> grid.draw(tg1)
> popViewport()
> grid.draw(rg1)
```

---

[*]The `rect` grob draws two rectangles: one thick and dark gray, one white and thin.

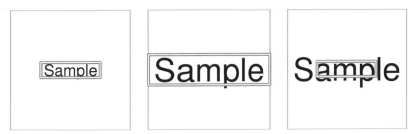

Figure 7.10
Calculating the size of a grob. In the left-hand panel, a `text` grob and a separate `rect` grob, the size of which is calculated to be the size of the `text` grob, are drawn together. In the middle panel, these objects are drawn together in a viewport with a larger font size, so they are both larger. In the right-hand panel, only the text is drawn in a viewport with a larger font size, so only the text is larger. The rectangle calculates the size of the text in a different font context.

A related issue arises with the use of grob *names* when creating a `"grobwidth"` or `"grobheight"` unit (see Section 6.3.2). The following code provides a simple example.

A `text` grob and two `rect` grobs are created, with the dimensions of both rectangles based upon the dimensions of the text. One rectangle, `rg1` (the gray one), uses the name `"tg1"` to refer to the text grob in the calls to `grobWidth()`, and `grobHeight()`. The other rectangle, `rg2` (the white one), just uses the `text` grob itself.

```
> tg1 <- textGrob("Sample", name="tg1")
> rg1 <- rectGrob(width=1.1*grobWidth("tg1"),
                  height=1.3*grobHeight("tg1"),
                  gp=gpar(col="gray60", lwd=3))
> rg2 <- rectGrob(width=1.1*grobWidth(tg1),
                  height=1.3*grobHeight(tg1),
                  gp=gpar(col="white"))
```

When these rectangles and text are initially drawn, both rectangles frame the text correctly (see the left panel of Figure 7.11).

```
> grid.draw(tg1)
> grid.draw(rg1)
> grid.draw(rg2)
```

However, if the `text` grob is modified, as shown below, only the rectangle `rg1` (the dark gray rectangle) will be updated to correspond to the new dimensions

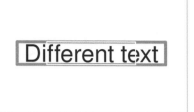

Figure 7.11
Grob dimensions by reference. In the left-hand panel there are three grobs: one
text grob and two **rect** grobs. The sizes of both **rect** grobs are calculated from the
text grob. The difference is that the white rectangle is related to the text by value
and the dark gray rectangle is related to the text by reference. The right-hand panel
shows what happens when the **text** grob is edited. Only the dark gray, by-reference,
rectangle gets resized.

of the text (see the right panel of Figure 7.11). The **rg1** rectangle, with its
reference to the grob name "**tg1**" will get the latest (modified) text grob, but
the **rg2** rectangle only has the original text grob to work with.

```
> grid.edit("tg1", grep=TRUE, global=TRUE,
            label="Different text")
```

With this approach, "**grobwidth**" and "**grobheight**" units are still evaluated
in the current geometric and graphical context, but in addition, only grobs
that have previously been drawn can be referred to. For example, drawing
the rectangle **rg1** before drawing the text **tg1** will not work because there is
no drawn grob named "**tg1**" from which a size can be calculated.

```
> grid.newpage()
> grid.draw(rg1)
```

Error in (function (name)   :
    grob 'tg1' not found

7.11.2   Calculating the positions of grobs

In addition to being able to query a grob about its dimensions, it is also
possible to query a grob about its location, using "**grobx**" and "**groby**" units,
or the **grobX()** and **grobY()** functions.

Locations are calculated relative to the current geometric and graphical context, just like widths and heights, so all of the warnings from the previous
section also apply here.

Figure 7.12
Calculating grob locations. The line segment is drawn from an explicit (x, y) start
location to an end location that is calculated using grobX() to give the left edge of
the box surrounding the text.

The grob locations are positions on the border of a grob, given by an angle
(relative to the "center" of the grob). The following code shows a simple
example usage (see Figure 7.12). A small dot is drawn on the left and a text
label, with a surrounding box, is drawn on the right. The box grob is named
"labelbox".

```
> grid.circle(.25, .5, r=unit(1, "mm"),
              gp=gpar(fill="black"))
> grid.text("A label", .75, .5)
> grid.rect(.75, .5,
            width=stringWidth("A label") + unit(2, "mm"),
            height=unit(1, "line"),
            name="labelbox")
```

A line segment, with an arrow, is now drawn between the dot and the left
edge of the box, using the grobX() function to determine the location of the
left edge of the box.

```
> grid.segments(.25, .5,
                grobX("labelbox", 180), .5,
                arrow=arrow(angle=15, type="closed"),
                gp=gpar(fill="black"))
```

The next example demonstrates a more complex use. This replicates an example from Figure 3.19 and demonstrates a possible use for "null" grobs.

First of all, two viewports are created, one in the top half of the page and one
in the bottom half.

```
> vptop <- viewport(width=.9, height=.4, y=.75,
                    name="vptop")
> vpbot <- viewport(width=.9, height=.4, y=.25,
                    name="vpbot")
> pushViewport(vptop)
> upViewport()
> pushViewport(vpbot)
> upViewport()
```

Now a rectangle and a line through some data are drawn in each viewport.

```
> grid.rect(vp="vptop")
> grid.lines(1:50/51, runif(50), vp="vptop")
> grid.rect(vp="vpbot")
> grid.lines(1:50/51, runif(50), vp="vpbot")
```

The next step does not draw anything, it just locates several null grobs at specific locations, two in the top viewport and two in the bottom viewport.

```
> grid.null(x=.2, y=.95, vp="vptop", name="tl")
> grid.null(x=.4, y=.95, vp="vptop", name="tr")
> grid.null(x=.2, y=.05, vp="vpbot", name="bl")
> grid.null(x=.4, y=.05, vp="vpbot", name="br")
```

Finally, a polygon is drawn that spans *both* viewports. The first two vertices of the polygon are calculated from the positions of the two null grobs in the top viewport and the second two vertices of the polygon are calculated from the positions of the two null grobs in the bottom viewport.

```
> grid.polygon(unit.c(grobX("tl", 0),
                      grobX("tr", 0),
                      grobX("br", 0),
                      grobX("bl", 0)),
              unit.c(grobY("tl", 0),
                      grobY("tr", 0),
                      grobY("br", 0),
                      grobY("bl", 0)),
              gp=gpar(col="gray", lwd=3))
```

The final result is shown in Figure 7.13.

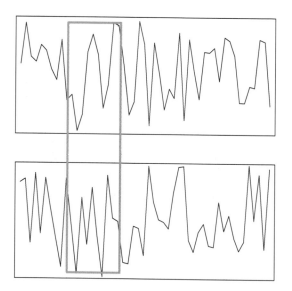

Figure 7.13
Calculating null grob locations. The two line plots are drawn in separate viewports. The thick gray rectangle is drawn relative to the locations of four null grobs, two of which are located in the top viewport and two of which are located in the bottom viewport.

7.12   Placing and packing grobs in frames

When drawing labels or legends on a plot, one of the difficult problems is determining sufficient margins for the labels or legends. The `"grobwidth"` and `"grobheight"` coordinate systems provide a way to determine the size of a grob and can be used to achieve this sort of arrangement of components by, for example, allocating appropriate regions within a layout.

The following code demonstrates this idea. First of all, some grobs are created to use as components of a scene. The first grob, `label`, is a simple `text` grob. The second grob, `gplot`, is a gTree containing a `rect` grob, a `lines` grob, and a `points` grob that provide a simple representation of time-series data. The `gplot` grob has a viewport in its `vp` component and the rectangle and lines are drawn within that viewport.

```
> label <- textGrob("A\nPlot\nLabel ",
                    x=0, just="left")
> x <- seq(0.1, 0.9, length=50)
> y <- runif(50, 0.1, 0.9)
> gplot <-
    gTree(
      children=gList(rectGrob(gp=gpar(col="gray60",
                                      fill="white")),
                    linesGrob(x, y),
                    pointsGrob(x, y, pch=16,
                              size=unit(1.5, "mm"))),
      vp=viewport(width=unit(1, "npc") - unit(5, "mm"),
                  height=unit(1, "npc") - unit(5, "mm")))
```

The next piece of code defines a layout with two columns. The second column of the layout has its width determined by the width of the `label` grob created above. The first column will take up whatever space is left over.

```
> layout <- grid.layout(1, 2,
                        widths=unit(c(1, 1),
                                    c("null", "grobwidth"),
                                    list(NULL, label)))
```

Now some drawing can occur. A viewport is pushed with the layout defined above, then the `label` grob is drawn in the second column of this layout, which is exactly the right width to contain the text, and the `gplot` gTree is drawn in the first column (see Figure 7.14).

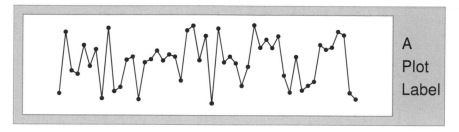

Figure 7.14
Packing grobs by hand. The scene was created using a frame object, into which the time-series plot (consisting of a rectangle, lines, and points) was packed. The text was then packed on the right-hand side, which meant that the time series plot was allocated less room in order to leave space for the text.

```
> pushViewport(viewport(layout=layout))
> pushViewport(viewport(layout.pos.col=2))
> grid.draw(label)
> popViewport()
> pushViewport(viewport(layout.pos.col=1))
> grid.draw(gplot)
> popViewport(2)
```

The **grid** package provides a set of functions that make it more convenient to arrange grobs like this so that they allow space for each other. The function `grid.frame()`, and its off-screen counterpart `frameGrob()`, produce a gTree with no children. Children are added to the frame using the `grid.pack()` function and the frame makes sure that enough space is allowed for the child when it is drawn. Using these functions, the previous example becomes simpler, as shown by the following code (the output is the same as Figure 7.14). The big difference is that there is no need to specify a layout as an appropriate layout is calculated automatically.

The first call creates an empty frame. The second call packs `gplot` into the frame; at this stage, `gplot` takes up the entire frame. The third call packs the text label on the right-hand side of the frame; enough space is made for the text label by reducing the space allowed for the rectangle.

```
> grid.frame(name="frame1")
> grid.pack("frame1", gplot)
> grid.pack("frame1", label, side="right")
```

There are many arguments to `grid.pack()` for specifying where to pack new grobs within a frame. There is also a **dynamic** argument to specify whether the frame should reallocate space if the grobs that have been packed in the frame are modified.

Unfortunately, packing grobs into a frame like this becomes quite slow as more grobs are packed, so it is most useful for very simple arrangements of grobs or for interactively constructing a scene. An alternative approach, which is a little more work, but still more convenient than dealing directly with pushing and popping viewports (and can be made dynamic like packing), is to *place* grobs within a frame that has a predefined layout. The following code demonstrates this approach. This time, the frame is initially created with the desired layout as defined above, then the `grid.place()` function is used to position grobs within specific cells of the frame layout.

```
> grid.frame(name="frame1", layout=layout)
> grid.place("frame1", gplot, col=1)
> grid.place("frame1", label, col=2)
```

7.12.1   Placing and packing off-screen

In the previous two examples, the screen is redrawn each time a grob is packed into the frame. An alternative is to create a frame and pack or place grobs within it off-screen and only draw the frame once it is complete. The following code demonstrates the use of the `frameGrob()` and `placeGrob()` functions to achieve the same end result as shown in Figure 7.14, doing all of the construction of the frame off-screen.

```
> fg <- frameGrob(layout=layout)
> fg <- placeGrob(fg, gplot, col=1)
> fg <- placeGrob(fg, label, col=2)
> grid.draw(fg)
```

The function `packGrob()` is the off-screen counterpart of `grid.pack()`.

7.13   Display lists

R's graphics engine maintains a display list, which is a record of all graphical output on a page, and this is used to redraw a scene if a page is resized (among other things; see Section 9.6). The output from both base and **grid** graphics functions is recorded on this display list.

The **grid** package also maintains its own separate display list, which is used for accessing grobs in the current scene and for redrawing the current scene after editing (i.e., after a call to `grid.edit()`). The **grid** display list can be replayed explicitly using the `grid.refresh()` function.

The **grid** display list can be disabled using `grid.display.list()`, which saves on **grid**'s memory usage, but disables **grid**'s ability to modify and redraw a scene. If the **grid** display list is disabled, the functions `grid.edit()`, `grid.get()`, `grid.add()`, and `grid.remove()` will no longer work.

It is possible to record **grid** output only on the **grid** display list with the `engine.display.list()` function, as shown by the following code. Redrawing will be slightly slower, but this avoids the memory cost of having output recorded on both the **grid** display list and the graphics engine display list.

```
> engine.display.list(FALSE)
```

This action only affects the recording of **grid** operations on the graphics engine display list; base graphics output is still recorded on the graphics engine display list.

7.14 Working with lattice grobs

The output from a **lattice** function is fundamentally just a collection of **grid** viewports and grobs. Section 6.8 described some examples of making use of the **grid** viewports that are set up by a **lattice** plot to add extra output. This section looks at some examples of working with the grobs that are created by a **lattice** plot.

The following code creates a **lattice** scatterplot to work with.

```
> xyplot(mpg ~ disp, mtcars)
```

The `grid.ls()` function shows the set of graphical primitives that have been created for this plot.

```
> grid.ls()
```

plot_01.background
plot_01.xlab
plot_01.ylab
plot_01.ticks.top.panel.1.1
plot_01.ticks.left.panel.1.1
plot_01.ticklabels.left.panel.1.1
plot_01.ticks.bottom.panel.1.1
plot_01.ticklabels.bottom.panel.1.1
plot_01.ticks.right.panel.1.1
plot_01.xyplot.points.panel.1.1
plot_01.border.panel.1.1

The grobs created by other people's functions will not necessarily provide useful names for all components that are drawn, but in this case, it is easy to spot which components provide the x-axis label and y-axis label for the plot.

The following code edits the axis labels, changes the font face to bold, and positions the labels at the ends of the axes (see Figure 7.15).

```
> grid.edit("[.]xlab$", grep=TRUE,
            label="Displacement",
            x=unit(1, "npc"), just="right",
            gp=gpar(fontface="bold"))
> grid.edit("[.]ylab$", grep=TRUE,
            label="Miles per Gallon",
            y=unit(1, "npc"), just="right",
            gp=gpar(fontface="bold"))
```

Other grob operations are also possible. For example, the following code removes the labels from the plot.

```
> grid.remove(".lab$", grep=TRUE, global=TRUE)
```

Finally, it is possible to group all of the grobs from a **lattice** plot together using `grid.grab()`. This creates a gTree that can then be used as a component in creating another picture.

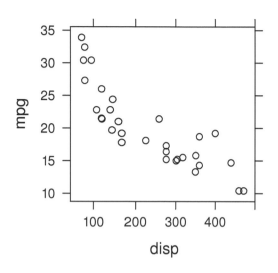

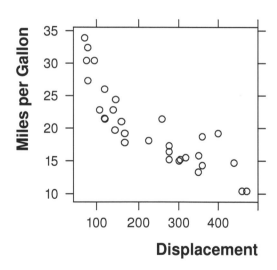

Figure 7.15
Editing the grobs in a **lattice** plot. The top plot is an initial scatterplot produced
using the **lattice** function `xyplot()`. The bottom plot shows the effect of editing
the **grid text** grobs that represent the labels on the plot (the labels are relocated
at the ends of the axes and are drawn in a monospace font).

7.15   Working with ggplot2 grobs

Like **lattice**, **ggplot2** creates lots of **grid** grobs when it draws a plot and these grobs can be manipulated using **grid** functions.

The following code uses **ggplot2** to create a scatterplot with a linear model line of best fit.

```
> ggplot(mtcars2, aes(x=disp, y=mpg)) +
      geom_point() +
      geom_smooth(method=lm)
```

The grobs generated by **ggplot2** are different from the grobs generated by **lattice** in several important ways. First of all, **ggplot2** only creates one single large gTree that only creates individual grobs as it is drawn. We can see this if we call **grid.ls()** for this **ggplot2**.

```
> grid.ls()
```

```
layout
```

If we want to access the individual grobs or viewports for the **ggplot2** plot, we must first call **grid.force()** (see Section 7.7).

```
> grid.force()
```

The output from **grid.ls()** now shows many individual grobs.

```
> grid.ls()
```

```
layout
  background.1-9-12-1
  panel.7-5-7-5
    grill.gTree.193
      panel.background..rect.184
      panel.grid.minor.y..polyline.186
      panel.grid.minor.x..polyline.188
      panel.grid.major.y..polyline.190
      panel.grid.major.x..polyline.192
    NULL
    geom_point.points.175
    geom_smooth.gTree.180
      geom_ribbon.polygon.177
      GRID.polyline.178
    NULL
    panel.border..zeroGrob.181
  spacer.8-6-8-6
  spacer.8-4-8-4
  spacer.6-6-6-6
  spacer.6-4-6-4
  axis-t.6-5-6-5
  axis-l.7-4-7-4
    axis.line.y.left..zeroGrob.206
    axis
      axis.1-1-1-1
        GRID.text.203
      axis.1-2-1-2
  axis-r.7-6-7-6
  axis-b.8-5-8-5
    axis.line.x.bottom..zeroGrob.199
    axis
      axis.1-1-1-1
      axis.2-1-2-1
        GRID.text.196
  xlab-t.5-5-5-5
  xlab-b.9-5-9-5
    GRID.text.210
  ylab-l.7-3-7-3
    GRID.text.213
  ylab-r.7-7-7-7
  subtitle.4-5-4-5
  title.3-5-3-5
  caption.10-5-10-5
  tag.2-2-2-2
```

The next big difference between **ggplot2** and **lattice** is that the grobs created by **ggplot2** are arranged in a hierarchy of grobs, which is visible in the indenting of the output from `grid.ls()`. This means that we may need to use gPaths to access individual grobs.

The final difference between **ggplot2** and **lattice** is that the **ggplot2** grob names are a little less meaningful, particularly the numeric suffixes. This means that we may need to use `grid.grep()` to find an appropriate grob name. The next code demonstrates using `grid.grep()` to find the viewport with `"panel"` in its name.

```
> panelvp <- grid.grep("panel", grobs=FALSE,
                          viewports=TRUE, grep=TRUE)
> panelvp
```

Now that we have forced the **ggplot2** plot grobs and viewports and we have determined the correct name of the main panel viewport, the following code navigates down to the plot region and queries the grob that represents the line of best fit, using `grobX()` and `grobY()`, to determine a location on the line. This location is used to draw an arrow that points from a text label to the line of best fit (see Figure 7.16).

```
> downViewport(panelvp)
> sline <- grid.get(gPath("smooth", "polyline"),
                      grep=TRUE)
> grid.segments(.7, .8,
                  grobX(sline, 45), grobY(sline, 45),
                  arrow=arrow(angle=10, type="closed"),
                  gp=gpar(fill="black"))
> grid.text("line of best fit", .71, .81,
              just=c("left", "bottom"))
```

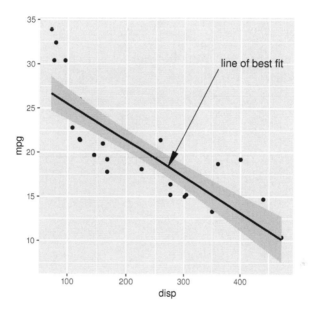

Figure 7.16
Working with **ggplot2** grobs. A **ggplot2** scatterplot is drawn and then a line is added with an end point that is calculated from the grob that represents the smooth line on the plot.

Chapter summary

As well as producing graphical output, all **grid** functions create grobs (graphical objects) that contain descriptions of what has been drawn. These grobs may be accessed, modified, and even removed, and the graphical output will be updated to reflect the changes.

There are also **grid** functions for creating grobs without producing any graphical output. A complete description of a plot can be produced by creating, modifying, and combining grobs off-screen.

A gTree is a grob that can have other grobs as its children. A gTree can be useful for grouping grobs and for providing a high-level interface to a group of grobs.

The **lattice** and **ggplot2** plotting functions generate large numbers of **grid** grobs. These grobs may be manipulated just like any other grobs to access, edit, and delete parts of a **ggplot2** or **lattice** plot.

8

Developing New Graphical Functions and Objects

Chapter preview

This chapter looks in depth at the task of writing graphical functions for others to use.

There are important guidelines for writing simple functions whose main purpose is to produce graphical output. There is an emphasis on making sure that other users can annotate the output produced by a function and that other users can make use of the function as a component in larger or more complex plots.

There is also a discussion on how to create a new class of graphical object. This is important for allowing users to edit output, to ask questions such as how much space a graphical object requires, and to be able to combine graphical objects together in a gTree.

This chapter addresses the issue of developing graphics functions for others to use. This will involve a discussion of some of the lower-level details of how **grid** works as well as some more abstract ideas of software design. A basic understanding of programming concepts is recommended, and the later sections assume an understanding of object-oriented concepts such as classes and methods.

Important low-level details of the **grid** graphics system and important design considerations are introduced in increasing levels of complexity to allow developers to construct simple graphics functions at first. Readers aiming to design a new **grid** graphical object should read the entire chapter.

Figure 8.1
An example of underlined text.

Although very complex **grid** functions and objects can be developed, (see, for example, the **gtable** package), this chapter will work with a very simple example. This is so that the functions and objects that we build are simple enough for the main ideas to be demonstrated clearly.

8.1   An example

Throughout this chapter we will develop code to generate very simple graphical output: text with an underline (see Figure 8.1).

The following code demonstrates one way we could produce this output using existing **grid** functions. Our aim is to reduce this code to a single function call.

```
> grid.text("underlined text", y=.5, just="bottom")
> w <- stringWidth("underlined text")
> grid.segments(unit(.5, "npc") - 0.5*w,
                unit(.5, "npc") - unit(1, "mm"),
                unit(.5, "npc") + 0.5*w,
                unit(.5, "npc") - unit(1, "mm"))
```

Although the output in Figure 8.1 is very basic, it provides examples of several key features that need to be addressed in much more complex output:

1. The output consists of more than one basic shape.

2. The placement of the output requires some calculation, and/or the components of the output are dependent on each other in some way.

```
1 textCorners <- function(x) {
2     list(xl=grobX(x, 180), xr=grobX(x, 0),
3           yb=grobY(x, 270), yt=grobY(x, 90))
4 }

6 grid.utext <- function(label, x=.5, y=.5, ...,
7                         name="utext") {
8     grid.text(label, x, y, ..., name=paste0(name, ".label"))
9     corners <- textCorners(paste0(name, ".label"))
10    grid.segments(corners$xl, corners$yb - unit(.2, "lines"),
11                  corners$xr, corners$yb - unit(.2, "lines"),
12                  gp=gpar(lex=get.gpar("cex")),
13                  name=paste0(name, ".underline"))
14 }
```

Figure 8.2
The `grid.utext()` function. This function draws underlined text.

In this case, there are two basic shapes, a piece of text and a line segment, and the constraints are that the line segment must be placed directly below the text and must be the same length as the text.

In solving this simple example, we will discuss several different approaches and solve a number of different issues that can arise when producing much more complex output.

8.2   Graphical functions

The simplest approach we can take is to write a graphics function just for its side effect of producing graphical output (i.e., using **grid** graphics functions as described in Chapter 6). Figure 8.2 provides code that defines a function `grid.utext()` for this purpose.

This conveniently encapsulates the work required to draw a line segment beneath a piece of text into a single function call, as shown below (see Figure 8.3).

```
> grid.utext("underlined text")
```

<u>underlined text</u>

Figure 8.3

An example of underlined text produced using the `grid.utext()` function.

8.2.1   Modularity

Although the `grid.utext()` function is very simple, the code in Figure 8.2 demonstrates an important general principle.

It is useful to organize code into small functions, each of which performs a well-defined job. With graphics functions, it is particularly important to separate out the calculations that underlie a graphic from the code that actually renders the graphic. In Figure 8.2 the function `textCorners()` calculates the left, right, bottom, and top locations of a text grob. The `grid.utext()` function draws text and a line segment and uses `textCorners()` to decide where to position the line segment so that it aligns with the text. This modular approach has the usual benefit of allowing the reuse of code, both by ourselves (we will reuse the `textCorners()` function several times in this chapter) and by others.

For more complex graphical output, this idea should be extended to organize code into functions that each produce separate components that can be combined into a larger graphic, just like this function combines existing graphical primitives.

8.2.2   Embeddable output

Another thing to consider when writing a **grid** function is the fact that all **grid** drawing occurs within the current viewport. This means that a function that produces **grid** output should be aware that it may be drawn in an area of any size and with any graphical parameter settings.

This is reflected in the code for the `grid.utext()` function on lines 10 and 11 where `"lines"` units are used to determine the vertical distance of the line segment from the text. This means that the distance will grow and shrink as the text size grows and shrinks. The following code demonstrates this idea by drawing underlined text within viewports with different `cex` settings.

Figure 8.4

A demonstration of the importance of using relative units within a **grid** function. This allows the function to perform well in any context. On the left, the
`grid.utext()` function is called within two viewports, one with a normal font (top)
and one with a small font (bottom). The output looks appropriate in both cases
because the distance between the text and the line segment is based on lines of text
and the line width is also relative. On the right, a variation of `grid.utext()` is
demonstrated where the distance between line and text is absolute (1mm); this does
not adjust to the different viewport contexts and consequently the resulting output
does not look as good.

The result is shown in Figure 8.4 alongside what would happen if we used an
absolute distance between the text and the line like 1mm instead.

```
> pushViewport(viewport(y=.5, height=.5, just="bottom",
                        gp=gpar(cex=1)))
> grid.utext("underlined text")
> popViewport()
> pushViewport(viewport(y=0, height=.5, just="bottom",
                        gp=gpar(cex=0.5)))
> grid.utext("underlined text")
> popViewport()
```

8.2.3   Editable output

Another important feature of the code for the `grid.utext()` function occurs
on lines 8 and 13, where the output that this function produces is named.
This ensures that others are able to edit the output from this function.

For example, in the following code we draw underlined text and then edit the
line segment to modify its width (and set the line end style to `"butt"`; see
Figure 8.5).

```
> grid.utext("underlined text")
> grid.edit("utext.underline", gp=gpar(lwd=3, lineend="butt"))
```

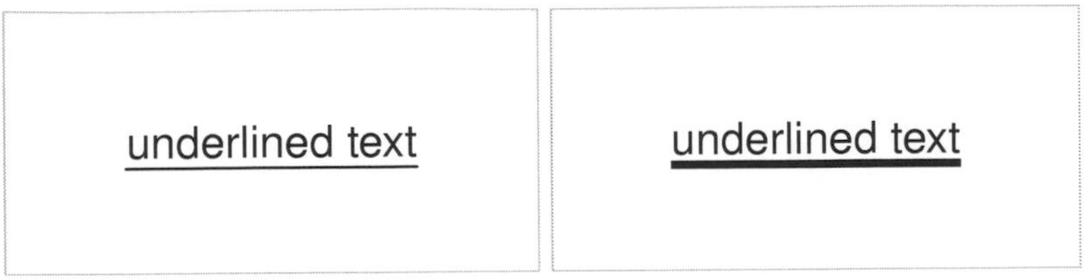

Figure 8.5
Underlined text from the `grid.utext()` function (left) is edited to modify the width of the underline (right).

8.2.4   Annotatable output

We have addressed the issue of adding the output from `grid.utext()` to other **grid** output in Section 8.2.2. Another issue is adding other **grid** output to the output from `grid.utext()`.

In order to demonstrate this point, we will consider an alternative implementation of underlined text that involves creating a **grid** viewport. This function is called `grid.utextvp()` and the code for it is shown in Figure 8.6.

One advantage of this function is that it can draw *rotated* underlined text, like that drawn by the following code (see Figure 8.7).

```
> grid.utextvp("underlined text", angle=20)
```

The important code within this function is the code that creates and pushes a viewport (line 9), with the name `"utextvp"` (line 4), and the code that calls `upViewport()` (line 15). The use of a viewport makes the code that does the drawing simpler (lines 10 to 14) because that drawing is relative to the viewport, which has already been positioned and sized appropriately for the text. The use of `upViewport()` is important because that means that the viewport will persist so that others can use it later.

The following code makes use of this feature to add a second underline beneath the text. First, we navigate down to the viewport that `utextvp()` created, and then we draw a line segment relative to that viewport (see Figure 8.8).

```
> downViewport("utextvp")
> grid.segments(0, unit(-.3, "lines"), 1, unit(-.3, "lines"))
```

```
1 utextvp <- function(label, x, y, ..., name="utextvp") {
2     w <- stringWidth(label)
3     viewport(x, y, width=w, height=unit(1, "lines"),
4             ..., name=name)
5 }

7 grid.utextvp <- function(label, x=.5, y=.5, ...,
8                          name="utext") {
9     pushViewport(utextvp(label, x, y, ...))
10     grid.text(label, y=0, just="bottom",
11             name=paste0(name, ".label"))
12     grid.segments(0, unit(-.2, "lines"),
13                   1, unit(-.2, "lines"),
14                   name=paste0(name, ".underline"))
15     upViewport()
16 }
```

Figure 8.6
The `grid.utextvp()` function. This function draws underlined text. It is an alternative to the `grid.utext()` function (Figure 8.2) that makes use of a **grid** viewport.

Figure 8.7
An example of underlined text produced using the `grid.utextvp()` function. This function can draw *rotated* underlined text (whereas the `grid.utext()` function cannot).

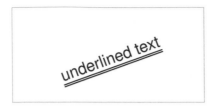

Figure 8.8
An example of annotating underlined text produced using the `grid.utextvp()` function (from Figure 8.7). The `grid.utextvp()` function creates a viewport for drawing the underlined text. This means that we can navigate back to that viewport and add further output, in this case a second underline.

8.3   Graphical objects

A properly written graphics function can be very useful if it can be reused in other plots and arbitrarily added to or modified as described in previous sections. There are, however, a number of benefits to be gained from instead creating a graphical object, or grob, to represent the output that your function produces.

In order to demonstrate a weakness of the graphical function approach, consider the following code, which edits the text that was produced by the `grid.utextvp()` function. This changes the text, but does not change the line segment, so the lengths of the two no longer match (see Figure 8.9).

```
> grid.edit("utext.label", label="le texte soulign\U00E9")
```

A graphical function is essentially an interface that allows us to provide a high-level description of an image and the function draws one or more low-level graphical shapes based on that description. However, all that is recorded on **grid**'s display list are the low-level graphical shapes.

In our example, the high-level description consists of the text that we want to draw, plus where to draw it, and the function `grid.utextvp()` calculates how to draw the line beneath the text (as well as the text itself). However, once these graphical shapes have been drawn, there is no high-level connection between them.

A graphical object is similar to a graphical function in that it provides a high-level interface for drawing low-level graphical shapes, but when we draw a graphical object, the high-level object is recorded on **grid**'s display list,

Figure 8.9
Underlined text from the `grid.utextvp()` function (Figure 8.7) has been edited to modify the text. The line segment is not affected so it no longer matches the text.

instead of or as well as the low-level shapes. This makes it possible to modify the high-level description and have the graphical object recalculate the low-level shapes (and redraw them).

In this section, we will develop a graphical object version of underlined text.

Defining new grobs involves working with classes and generic functions. This section assumes a familiarity with the basic ideas of object-oriented programming and its implementation in S3 classes and methods.

8.3.1 Defining a static grob

The simplest graphical object that we can create is a gTree with explicit children; a *static* graphical object. The code shown in Figure 8.10 defines a `utextStatic()` function for this purpose.

The most important part of this function is the call to the `gTree()` function (line 19). This creates a gTree that records the high-level description of the graphical object (the text and where to draw it). The `cl` argument is used to specify that the object created by this function has the class `"utextStatic"`. This will allow us to define methods specific to `"utextStatic"` grobs. The children of the gTree are the low-level grobs that will actually be drawn; a text grob and a line segment grob. These are created in a separate function called `utextChildren()` so that we can reuse that function in later examples.

The following code creates a `"utextStatic"` grob and then draws it (see Figure 8.11). The default drawing behaviour of a gTree is to draw all of its children.

```
> ug <- utextStatic("underlined text")
> grid.draw(ug)
```

```
 1 utextChildren <- function(label, x, y, just, name) {
 2     t <- textGrob(label, x, y, just=just,
 3                     name=paste0(name, ".label"))
 4     corners <- textCorners(t)
 5     s <- segmentsGrob(corners$xl,
 6                         corners$yb - unit(.2, "lines"),
 7                         corners$xr,
 8                         corners$yb - unit(.2, "lines"),
 9                         name=paste0(name, ".underline"))
10     gList(t, s)
11 }

13 utextStatic <- function(label,
14                          x=.5, y=.5, default.units="npc",
15                          just="centre", name="utext") {
16     if (!is.unit(x)) x <- unit(x, default.units)
17     if (!is.unit(y)) y <- unit(y, default.units)
18     kids <- utextChildren(label, x, y, just, name)
19     gTree(label=label, x=x, y=y, just=just,
20           children=kids, cl="utextStatic", name=name)
21 }
```

Figure 8.10
The `utextStatic()` function. This function creates a gTree object representing underlined text. It is a graphical object alternative to the `grid.utext()` graphical function (Figure 8.2).

Figure 8.11
An example of underlined text produced using the `utextStatic()` function. The `utextStatic()` function creates a grob representing underlined text, but does not draw it. The resulting grob must be drawn by calling the `grid.draw()` function.

```
1 editDetails.utextStatic <- function(x, specs) {
2     if (any(names(specs) %in%
3         c("label", "x", "y", "just"))) {
4         kids <- utextChildren(x$label, x$x, x$y,
5                                 x$just, x$name)
6         x <- setChildren(x, kids)
7     }
8     x
9 }
```

Figure 8.12
The `editDetails.utextStatic()` function. This function allows us to sensibly modify the high-level description of a gTree object by recreating the low-level children
of the object.

8.3.2   Editable grobs

What we have created is a graphical object that contains a high-level description plus low-level components (text and a line segment).

```
> grid.ls()
```

utext
  utext.label
  utext.underline

If we want to edit the high-level description of a grob, we must define an
`editDetails()` method for the graphical object. Figure 8.12 shows a method
for `"utextStatic"` grobs. The important role of this function is to recreate
the children of the gTree if the high-level description has been modified.

Once we have defined an `editDetails()` method, we can modify the high-level
description and the low-level components are automatically recreated. The
following code modifies the label of the text and the underline is automatically
resized (see Figure 8.13).*

```
> grid.edit("utext", label="le texte soulign\U00E9")
```

Because we also have the low-level components, we can modify those to make
changes that the high-level interface does not allow. The following code modifies the line style of the underline (see Figure 8.14).

---

*The UNICODE escape sequence `\U00E9` is used to specify an e-acute character.

le texte souligné

Figure 8.13
Underlined text from the `utextStatic()` function (Figure 8.11) has been edited to
modify the `label` (part of the high-level description). The `editDetails()` method
ensures that the children of the graphical object are recreated, so the underline still
matches the text.

le texte souligné

Figure 8.14
Underlined text from the `utextStatic()` function (Figure 8.13) has been edited to
modify the underline segment line type (one of the low-level components). This
shows fine tuning of low-level details that are not controlled by the high-level description.

```
> grid.edit("utext.underline", gp=gpar(lty="dashed"))
```

8.3.3 Defining a static grob with drawing context

When we developed a graphical function in Section 8.2, we considered two
approaches: one that just drew text and a line segment, and another that
pushed a viewport and then drew text and a line segment within that viewport.
In this section, we will look at a static graphical object analogue of the second
approach.

A static graphical object can also have a viewport within which its children
are drawn. Figure 8.15 shows code for a `utextvpStatic()` function that
implements this approach. The difference from the `utextStatic()` (Figure
8.10) is that, as well as creating children for the gTree (line 18), we create a
viewport for the children to be drawn within (line 19). The viewport becomes
the `childrenvp` component of the gTree (line 22).

As with the `utextStatic()` function, we define an `editDetails()` method so that changes to the high-level description of the graphical object result in the children, and the viewport that the children are drawn within, being recreated.

The following code creates a `"utextvpStatic"` grob (at an angle of 20 degrees) and then draws it (see Figure 8.16).

```
> ug <- utextvpStatic("underlined text", angle=20)
> grid.draw(ug)
```

What we have created is a graphical object that contains a high-level description with a viewport plus low-level components (text and a line segment) that are drawn within that viewport.

```
> grid.ls(viewports=TRUE, fullNames=TRUE)
```

```
viewport[ROOT]
  utextvpStatic[utext]
    viewport[utext.vp]
      upViewport[1]
    downViewport[utext.vp]
      text[utext.label]
      upViewport[1]
    downViewport[utext.vp]
      segments[utext.underline]
      upViewport[1]
```

Because of the `editDetails()` method, we can modify the high-level description and the low-level components are automatically recreated (see Figure 8.17).

```
> grid.edit("utext", label="le texte soulign\U00E9")
```

Because we also have the low-level components, we can modify those to make changes that the high-level interface does not allow (see Figure 8.18).

```
> grid.edit("utext.underline", gp=gpar(lty="dashed"))
```

```
1 utextvpChildren <- function(label, name) {
2     t <- textGrob(label, y=0, just="bottom",
3                     vp=paste0(name, ".vp"),
4                     name=paste0(name, ".label"))
5     s <- segmentsGrob(0, unit(-.2, "lines"),
6                         1, unit(-.2, "lines"),
7                         vp=paste0(name, ".vp"),
8                         name=paste0(name, ".underline"))
9     gList(t, s)
10 }

12 utextvpStatic <- function(label, x=.5, y=.5,
13                             default.units="npc",
14                             angle=0, just="centre",
15                             name="utext") {
16     if (!is.unit(x)) x <- unit(x, default.units)
17     if (!is.unit(y)) y <- unit(y, default.units)
18     kids <- utextvpChildren(label, name)
19     kidsvp <- utextvp(label, x, y, just=just, angle=angle,
20                         name=paste0(name, ".vp"))
21     gTree(label=label, x=x, y=y, just=just, angle=angle,
22             children=kids, childrenvp=kidsvp,
23             cl="utextvpStatic", name=name)
24 }

26 editDetails.utextvpStatic <- function(x, specs) {
27     if (any(names(specs) %in%
28             c("label", "x", "y", "just", "angle"))) {
29         kids <- utextvpChildren(x$label, x$name)
30         kidsvp <- utextvp(x$label, x$x, x$y,
31                             just=x$just, angle=x$angle,
32                             name=paste0(x$name, ".vp"))
33         x$childrenvp <- kidsvp
34         x <- setChildren(x, kids)
35     }
36     x
37 }
```

Figure 8.15
The `utextvpStatic()` function. This function creates a gTree object representing underlined text. It is a graphical object alternative to the `grid.utextvp()` graphical function (Figures 8.6).

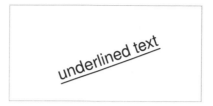

Figure 8.16
An example of underlined text produced using the `utextvpStatic()` function. The `utextvpStatic()` function creates a grob representing underlined text, but does not draw it. The resulting grob must be drawn by calling the `grid.draw()` function.

Figure 8.17
Underlined text from the `utextvptree()` function (Figure 8.16) has been edited to modify the `label` (part of the high-level description). The `editDetails()` method ensures that the children of the graphical object, and the viewport that the children are drawn within, are recreated.

Figure 8.18
Underlined text from the `utextvptree()` function (Figure 8.17) has been edited to modify the underline segment line type (one of the low-level components). This shows fine tuning of low-level details that are not controlled by the high-level description.

8.3.4   Defining a dynamic grob

It is not always possible to know, when a grob is created, exactly what its children should be. For example, an `"xaxis"` with `at=NULL` can only determine its children (the tick marks) when it is drawn, which is when it knows what viewport, and hence what `"native"` coordinate system it is being drawn within (see Section 7.7). This section looks at developing a graphical object that can create its children on-the-fly as it is drawn; a *dynamic* graphical object.

Figure 8.19 shows code that defines two functions. The first, `utextDynamic()`, is the function that creates a graphical object. This function does no drawing; it only creates a gTree object that contains the high-level description of the underlined text (the text to draw and where to draw it). The most important part of this function is the call to the `gTree()` function to create a gTree object (line 6). The `cl` argument is used to specify that the object created by this function has the class `"utextDynamic"`. This will allow us to define methods specific to `"utextDynamic"` grobs.

If we compare this `utextDynamic()` function to the previous `utextStatic()` function (Figure 8.10), the major difference is that `utextDynamic()` creates a gTree with no children. This is because, for a dynamic grob, the children are created when the `"utextDynamic"` grob is drawn. That is the purpose of the second function.

The second function is a method for the `makeContent()` generic function (for objects of class `"utextDynamic"`). This function creates the low-level shapes that are the "children" of the gTree (line 11), and then calls the `setChildren()` function to add these grobs as children of the `"utextDynamic"` gTree (line 13). The result of this function is the modified `"utextDynamic"` gTree.

The following code makes use of these functions to create a `"utextDynamic"` grob and then draws it by calling the `grid.draw()` function (see Figure 8.20).

```
> ug <- utextDynamic("underlined text")
> grid.draw(ug)
```

The output from drawing a `"utextDynamic"` grob is just the same as from calling the `grid.utext()` function and just the same as drawing a `"utextStatic"` grob. However, the **grid** display list is very different because the only object recorded is the `"utextDynamic"` grob, not the individual text grob or segments grob.

```
> grid.ls()

utext
```

```
1 utextDynamic <- function(label,
2                             x=.5, y=.5, default.units="npc",
3                             just="centre", name="utext") {
4     if (!is.unit(x)) x <- unit(x, default.units)
5     if (!is.unit(y)) y <- unit(y, default.units)
6     gTree(label=label, x=x, y=y, just=just,
7           cl="utextDynamic", name=name)
8 }

10 makeContent.utextDynamic <- function(x) {
11    kids <- utextChildren(x$label, x$x, x$y,
12                             just=x$just, x$name)
13    setChildren(x, kids)
14 }
```

Figure 8.19
The `utextDynamic()` function. This function creates a gTree object representing underlined text. It is a dynamic version of the `utext()` function from Figure 8.1; this grob creates its contents as it is drawn, whereas `utext()` creates its contents when it is created.

Figure 8.20
An example of underlined text produced using the `utextDynamic()` function. The `utextDynamic()` function creates a grob representing underlined text, but does not draw it. The resulting grob must be drawn by calling the `grid.draw()` function.

underlined text

Figure 8.21
Underlined text from the `utextDynamic()` function (Figure 8.20) has been edited to modify the `gp` settings for the overall `"utextDynamic"` gTree. Both children of the gTree, the label and the line segment, are affected by the graphical context of their parent gTree, so they both turn gray.

This demonstrates the idea that, with a dynamic graphical object, what gets recorded on the **grid** display list is just a high-level description. By comparison, with a graphical function, only low-level shapes get recorded on the display list, and, with a static graphical function, both the high-level description and the low-level shapes are recorded.

We gain several benefits from retaining the high-level description. One benefit arises from the fact that a gTree provides a graphical context for its children (see page 6.4); specifically, the `gp` settings for a gTree become the default `gp` settings for its children. This is demonstrated by the following code: we change the colour of the `"utextDynamic"` grob to gray and both the text and the underline turn gray as a result (see Figure 8.21).

```
> grid.edit("utext", gp=gpar(col="grey"))
```

Another benefit of having the high-level `"utextDynamic"` gTree on the display list is that we can edit that high-level description and the low-level shapes will be redrawn. This is demonstrated in the following code: we modify the label in the high-level gTree and both the text and the underline adjust for the new label (see Figure 8.22).

```
> grid.edit("utext", label="le texte soulign\U00E9")
```

Unlike a static graphical object (e.g., a `"utextStatic"` object), we do not have to define an `editDetails()` method. The children of a `"utextDynamic"` grob are always recreated whenever the `"utextDynamic"` grob is drawn.

le texte souligné

Figure 8.22
Underlined text from the `utextDynamic()` function (Figure 8.21) has been edited
to modify the label of the high-level `"utextDynamic"` gTree. Both children of the
gTree, the label and the line segment, are redrawn to reflect the new label.

8.3.5   Forcing grobs

The main downside to working with this high-level dynamic interface is that
we lose access to the low-level shapes. The display list only contains the high-
level gTree; it does not record the text grob or the segment grob. This means
that we cannot directly access the individual shapes, like we did in Figure 8.5.

```
> grid.get("utext.label")
```

NULL

However, we can make the low-level shapes available by calling the function
`grid.force()`. This function adds the children of a gTree to the **grid** display
list, as shown below.

```
> grid.force()
```

utext
  utext.label
  utext.underline

We can now access the individual low-level shapes. For example, in the following code, we increase the width of the underline (see Figure 8.23).

```
> grid.edit("utext.underline", gp=gpar(lwd=3))
```

le texte souligné

Figure 8.23
Underlined text from the `utextDynamic()` function (Figure 8.20) has been drawn,
then the low-level shapes have been made available by calling `grid.force()`, then
the underline segments grob has been edited to make the line thicker.

8.3.6   Reverting grobs

The downside to calling `grid.force()` is that we no longer have access to the
high-level gTree interface (changes to the gTree no longer recreate the children
of the gTree). There is a `grid.revert()` function that removes the children
of the gTree from the display list and restores the high-level interface, but
of course that means that direct changes to the low-level children are lost.
In other words, with a dynamic grob, we cannot have both high-level and
low-level access at the same time.

8.3.7   Defining a dynamic grob with drawing context

When we developed a graphical function in Section 8.2, we considered two
approaches: one which just drew text and a line segment and another that
pushed a viewport and then drew text and a line segment within that viewport.
In this section, we will look at a dynamic graphical object analogue of the
second approach.

We have seen that one thing we must do when we create a dynamic graphical
object is define a `makeContent()` method. The purpose of that is to create
the low-level children of the graphical object. In addition to that, we may
also want to create a viewport for the children of the graphical object. The
`makeContext()` generic function allows us to do that.

Figure 8.24 shows the code for a new graphical object that represents underlined text. This has a main function, `utextvpDynamic()` that creates a gTree
with the class `"utextvpDynamic"` (line 7). There is also a `makeContent()`
method to create low-level children for the gTree, like there was for the
`"utextDynamic"` graphical object, and in addition there is a `makeContext()`
method. A `makeContext()` method is similar to a `makeContent()` method,
except that it creates a *viewport* and adds it to the gTree (rather than creating

```
1 utextvpDynamic <- function(label,
2                                 x=.5, y=.5, default.units="npc",
3                                 just="centre", angle=0,
4                                 name="utext") {
5      if (!is.unit(x)) x <- unit(x, default.units)
6      if (!is.unit(y)) y <- unit(y, default.units)
7      gTree(label=label, x=x, y=y, just=just, angle=angle,
8            cl="utextvpDynamic", name=name)
9 }

11 makeContext.utextvpDynamic <- function(x) {
12     x$childrenvp <- utextvp(x$label, x$x, x$y,
13                             just=x$just, angle=x$angle,
14                             name=paste0(x$name, ".vp"))
15     x
16 }

18 makeContent.utextvpDynamic <- function(x) {
19     kids <- utextvpChildren(x$label, x$name)
20     setChildren(x, kids)
21 }
```

Figure 8.24
The utextvpDynamic() function. This function creates a gTree object representing underlined text. It is an alternative to the utextDynamic() function (Figure 8.19).

and adding grobs).

With these functions defined, it is possible to create a "utextvpDynamic" grob and draw it (see Figure 8.25). The advantage of utextvpDynamic() over utextDynamic() is that the former can draw underlined text at an angle.

```
> ug <- utextvpDynamic("underlined text", angle=20)
> grid.draw(ug)
```

Although the low-level children of a "utextvpDynamic" grob are not recorded on the **grid** display list, the viewports that are created in a makeContext() method do get recorded.

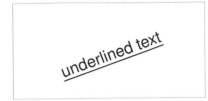

Figure 8.25
An example of underlined text produced using the `utextvpDynamic()` function. The
`utextvpDynamic()` function creates a grob representing underlined text, but does not
draw it. The resulting grob must be drawn by calling the `grid.draw()` function.

Figure 8.26
Underlined text from the `utextvpDynamic()` function (Figure 8.25) has been drawn,
then we have navigated down to the viewport that was used to draw the underlined
text and added a second underline.

```
> grid.ls(viewports=TRUE, fullNames=TRUE)
```

```
viewport[ROOT]
  utextvpDynamic[utext]
    viewport[utext.vp]
      upViewport[1]
```

In the following code, we navigate down to the viewport within which the
low-level shapes were drawn and add another underline (see Figure 8.26).

```
> downViewport("utext.vp")
> grid.segments(0, unit(-.3, "lines"), 1, unit(-.3, "lines"))
```

8.3.8   Querying graphical objects

A final advantage of creating a graphical object compared to a graphical
function is that we provide the opportunity for other code to query our

```
1 xDetails.utextvpDynamic <- function(x, theta) {
2     h <- unit(1, "npc") + unit(.2, "lines")
3     grobX(rectGrob(height=h, y=1, just="top",
4                    vp=paste0(x$name, ".vp")), theta)
5 }

7 yDetails.utextvpDynamic <- function(x, theta) {
8     h <- unit(1, "npc") + unit(.2, "lines")
9     grobY(rectGrob(height=h, y=1, just="top",
10                   vp=paste0(x$name, ".vp")), theta)
11 }
```

Figure 8.27
xDetails() and yDetails() methods for querying the edge locations of "utextvp-
Dynamic" grobs.

graphical object through the functions grobX(), grobY(), grobWidth(), and
grobHeight().

We can provide useful results to those functions for our graphical object by
defining methods for a set of generic functions: xDetails(), yDetails(),
widthDetails(), and heightDetails().

Figure 8.27 shows code for an xDetails() method and a yDetails() method
for "utextvpDynamic" grobs. One important feature of these functions is
that they create a simple grob, in this case a rectangle, and then call grobX()
(or grobY()) on that simple grob. In other words, they make use of existing
methods as much as possible. The second important feature of these functions
is that the simple grobs that they create are given vp values to reflect the fact
that they are relative to the childrenvp of the "utextvpDynamic" grob.

The following code makes use of these new methods to draw a line from a dot
in the top-left corner of the current viewport to the bottom-left corner of a
"utextvpDynamic" grob. The result is shown in Figure 8.28.

```
> ug <- utextvpDynamic("underlined text")
> grid.draw(ug)
> grid.circle(.1, .8, r=unit(1, "mm"), gp=gpar(fill="black"))
> grid.segments(.1, .8,
                grobX("utext", 180), grobY("utext", 270))
```

Figure 8.28
Underlined text from the `utextvpDynamic()` function has been drawn, which produces a `"utextvpDynamic"` grob, then a dot has been drawn at top-left, then the
`"utextvpDynamic"` grob has been queried for its bottom-left corner and a line has
been drawn from the dot to that corner.

8.3.9   Summary of graphical object methods

Defining the behavior for a new graphical object requires writing one or more
methods for the standard **grid** generic functions:

- **Always** write a constructor function for the class to generate a gTree
  containing the description of what to draw.
  For static grobs, the constructor should also create the children of the
  gTree and any viewports for the children to be drawn within.
- For dynamic grobs, **always** write a `makeContent()` method to create
  low-level children for the object (otherwise nothing will be drawn).
- For dynamic grobs, **sometimes** write a `makeContext()` method if drawing graphical object involves pushing viewports.
- **Sometimes** write `xDetails()` and `yDetails()` methods if the boundary of the graphical output can be sensibly determined.
- **Sometimes** write `widthDetails()` and `heightDetails()` methods if
  the size of the graphical output can be sensibly determined.

8.3.10   Calculations during drawing

With **grid** units and layouts, it is possible to specify quite complex arrangements of output in a "declarative" manner. For example, the idea that a particular region should be square (have an aspect ratio of 1) can be expressed
at a high level, by specifying both width and height as `unit(1, "snpc")`,
and the system will ensure that this occurs. There is no need to calculate the
physical dimensions of the current viewport and from those determine how to
make a square region.

It is, however, sometimes necessary to perform calculations by hand. For
example, consider the problem of splitting text into several lines based on

```
 1 splitString <- function(text) {
 2   strings <- strsplit(text, " ")[[1]]
 3   if (length(strings) < 2)
 4     return(text)
 5   newstring <- strings[1]
 6   linewidth <- stringWidth(newstring)
 7   gapwidth <- stringWidth(" ")
 8   availwidth <-
 9     convertWidth(unit(1, "npc"),
10                   "in", valueOnly=TRUE)
11   for (i in 2:length(strings)) {
12     width <- stringWidth(strings[i])
13     if (convertWidth(linewidth + gapwidth + width,
14                     "in", valueOnly=TRUE) <
15         availwidth) {
16       sep <- " "
17       linewidth <- linewidth + gapwidth + width
18     } else {
19       sep <- "\n"
20       linewidth <- width
21     }
22     newstring <- paste(newstring, strings[i], sep=sep)
23   }
24   newstring
25 }
```

Figure 8.29
A `splitString()` function. This function takes a piece of text and splits it into multiple lines so that the text will fit (horizontally) within the current viewport. Validation checks (e.g., whether **strings** is a character vector of length at least 2) have not been included.

the width of the available space. The code in Figure 8.29 defines a function, `splitString()`, to perform this operation (in a very simple-minded way). The important part of this function is the use of the `convertWidth()` function to obtain the size of the current line of text in inches (line 13) for comparison with the size of the current viewport in inches (lines 8 to 10).

The following code uses the `splitString()` function to draw some text within the current viewport (see the left-hand panel in Figure 8.30).

```
> text <- "The quick brown fox jumps over the lazy dog."
> grid.text(splitString(text),
            x=0, y=1, just=c("left", "top"))
```

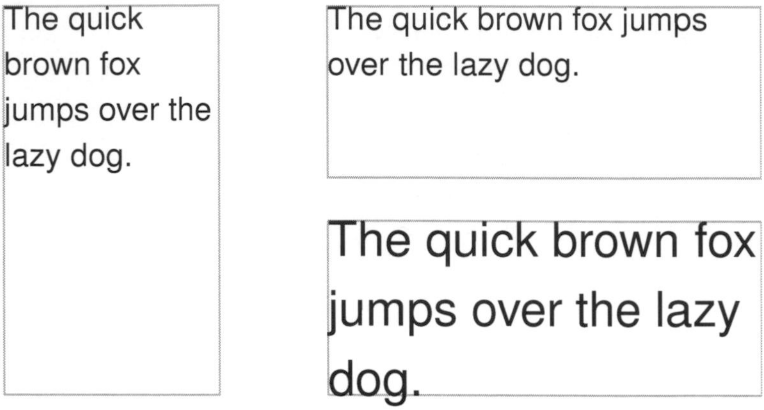

The quick
brown fox
jumps over the
lazy dog.

The quick brown fox jumps
over the lazy dog.

The quick brown fox
jumps over the lazy
dog.

Figure 8.30
Performing calculations before drawing. If the drawing of a grob depends on calculations (in this case, calculations to split text into multiple lines to fit horizontally
within the current viewport), the calculations should be included within a make-
Content() method. This means that the calculations will be rerun if the device is
resized (left panel versus top-right panel) or if the grob is edited to make the font
size larger (top-right panel versus bottom-right panel).

There is a problem with the above code. If it is used to draw into a window
and then the window is resized, the calculations are not rerun and the line
splitting becomes incorrect.

The issue is that only drawing actions are recorded on the display list, not any
calculations leading up to the drawing. Anything that works off the display
list (like redrawing after a resize) only reruns drawing actions.

There are two solutions to this problem. One solution rests on the fact that
all code within a makeContent() method (or a makeContext() method) is
captured on the graphics engine display list. The code in Figure 8.31 uses
this fact to create a "splitText" grob with a makeContent() method that
performs the calculations.

A splitText grob will recalculate the line breaks when a window is resized
(see the top-right panel of Figure 8.30).

```
> splitText <- splitTextGrob(text, name="splitText")
> grid.draw(splitText)
```

Another advantage of creating a grob with a makeContent() method is that
it is possible to edit the grob and have the calculations updated (see the
bottom-right panel of Figure 8.30).

```
1 splitTextGrob <- function(text, ...) {
2     gTree(text=text, cl="splitText", ...)
3 }

5 makeContent.splitText <- function(x) {
6     setChildren(x, gList(textGrob(splitString(x$text),
7                                   x=0, y=1,
8                                   just=c("left", "top"))))
9 }
```

Figure 8.31
A "splitText" grob. The `makeContent()` method for the class recalculates where
to place line breaks in the text, based on the current viewport size.

```
> grid.edit("splitText", gp=gpar(cex=1.5))
```

The other way to encapsulate calculations with drawing operations is to use
the `grid.delay()` function, as shown by the following code.

```
> grid.delay({
            grid.text(splitString(text),
                      x=0, y=1, just=c("left", "top"))
            },
            list(text=text))
```

This is convenient for writing code purely for its side effect (i.e., without having
to deal explicitly with grobs), but it provides less control over the design of
the object that is created. There is also a `delayGrob()` function that simply
creates a grob encapsulating the calculations and drawing operations without
drawing anything.

8.3.11 Avoiding argument explosion

Very complex or high-level graphics functions and objects are usually composed of several lower-level elements, which in turn may be composed of several even-lower-level elements. For example, a scatterplot matrix is composed
of several scatterplots and each scatterplot contains axes, labels, and data
symbols.

Ideally, it should be possible to control any aspect of a graphical scene. In
terms of writing code, this means that an argument or component should be

supplied to allow the user to specify a customized value for any parameter of the scene.

At the level of graphical primitives, parameters consist of such things as the locations of lines, the color of lines, and the line thickness. At a higher level, for example for axes, there are higher-level parameters, such as where to place tick marks, but it is also desirable to still be able to control the individual elements of the axis.

It is tempting to simply provide arguments for the elements of an axis as arguments of the axis itself. An example is where an axis could have a `rot` argument to specify the angle of rotation of the tick mark labels, but this approach quickly runs into difficulties. For one thing, ambiguities can easily arise. If an axis had an overall label, it is unclear whether the `rot` argument would apply to the tick mark labels or to the overall label. Another problem is that as elements become more complex, the number of parameters required for all subelements grows alarmingly. Consider the number of separate arguments required to individually specify the angle of rotation for tick mark labels on all scatterplots within a scatterplot matrix!

The **grid** package provides several features that can help to solve this problem. The functions `grid.edit()` and `editGrob()` (see Section 7.1) make it possible to access the lower-level elements of an object using a gPath. For example, in the following code, an x-axis is created and then the labels on the tick marks are rotated by editing the `rot` component of the `text` grob called `"labels"` that is a child of the `xaxis` grob.

```
> grid.xaxis(at=1:3/4, name="xaxis1")
> grid.edit("labels", rot=45)
```

More complex is the case where a grob calculates its children on the fly. This typically occurs when a grob has no permanent children to access via a gPath and this will often correspond to a grob that has a `makeContent()` method.

This problem can be solved by calling the function `grid.force()`, which makes the children of a grob visible on the display list.

```
> grid.xaxis(name="xaxis1")
> grid.force()
> grid.edit("labels", rot=45)
```

8.4   Mixing graphical functions and graphical objects

This chapter has addressed two main ways in which to develop new graphical functionality: as a graphics function, purely for the side effect of producing output (see Section 8.2); and as a graphical object (Section 8.3). There has also been an emphasis on producing reusable graphical elements, a corollary of which is that existing graphical elements should be used where possible in the construction of new graphical elements.

There is no way to force other developers to create graphical objects rather than graphical functions, so it is necessary to be able to make use of both existing functions and existing objects whether we are constructing a new function or a new object.

In order to discuss each of the four possible situations (new functions from existing functions, new functions from existing grobs, new grobs from existing functions, and new grobs from existing grobs), the following paragraphs consider the simple case of drawing a "face," which consists of a rectangle for the border, two circles for eyes, and a line for the mouth (see Figure 8.32 for examples).

Defining a new graphics function is straightforward whether using existing graphics functions or existing graphical objects. Figure 8.33 defines two new graphical functions to draw a face. The function `faceA()` demonstrates the most straightforward case of a graphics function that includes calls to other graphics functions to produce output (lines 4 to 6). The function `faceB()` shows a graphics function making use of existing graphical objects, which is done by just passing the result of the object constructor functions to the function `grid.draw()` (lines 13 to 15).

Developing a new graphical object can be a bit trickier, but there are several tools to help out. Figure 8.34 defines two functions for creating a graphical object to represent a face. The function `faceC()` represents the simplest case, where a gTree is built from existing graphical objects, by just creating the appropriate objects as children of the gTree (lines 5 to 10).

Function `faceD()` demonstrates the harder problem of creating a new graphical object using only existing graphics *functions*. In this case, a solution is to capture the output of the graphics function as a gTree with a call to the `grid.grabExpr()` function.

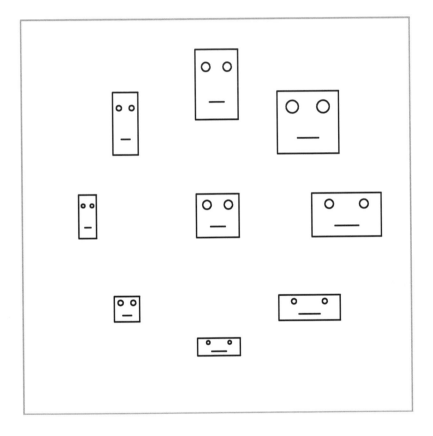

Figure 8.32
Drawing faces. Examples of the output that could be produced using the graphical
functions and graphical objects defined in Figures 8.33 and 8.34.

```
1 faceA <- function(x, y, width, height) {
2   pushViewport(viewport(x=x, y=y,
3                         width=width, height=height))
4   grid.rect()
5   grid.circle(x=c(0.25, 0.75), y=0.75, r=0.1)
6   grid.lines(x=c(0.33, 0.67), y=0.25)
7   popViewport()
8 }

10 faceB <- function(x, y, width, height) {
11   pushViewport(viewport(x=x, y=y,
12                         width=width, height=height))
13   grid.draw(rectGrob())
14   grid.draw(circleGrob(x=c(0.25, 0.75), y=0.75, r=0.1))
15   grid.draw(linesGrob(x=c(0.33, 0.67), y=0.25))
16   popViewport()
17 }
```

Figure 8.33
Some face functions. Some different ways to implement a new graphical function to draw a "face." The function faceA() makes use of existing graphical functions. The function faceB() makes use of existing graphical objects.

```
 1 faceC <- function(x, y, width, height) {
 2     gTree(childrenvp=viewport(x=x, y=y,
 3                               width=width, height=height,
 4                               name="face"),
 5           children=gList(rectGrob(vp="face"),
 6                          circleGrob(x=c(0.25, 0.75),
 7                                     y=0.75, r=0.1,
 8                                     vp="face"),
 9                          linesGrob(x=c(0.33, 0.67), y=0.25,
10                                    vp="face")))
11 }

13 faceD <- function(x, y, width, height) {
14     grid.grabExpr({
15                 pushViewport(viewport(x=x, y=y,
16                                       width=width,
17                                       height=height))
18               grid.rect()
19               grid.circle(x=c(0.25, 0.75),
20                           y=0.75, r=0.1)
21               grid.lines(x=c(0.33, 0.67), y=0.25)
22               popViewport()
23             })
24 }
```

Figure 8.34
Some face objects. Some different ways to implement a new graphical object to represent a "face." The function faceC() makes use of existing graphical objects. The function faceD() makes use of existing graphics functions by capturing their output as a gTree.

8.5   Debugging grid

One of the difficulties of working with **grid** is relating our R code to what we can see on screen. The R code describes what to draw, but when things go wrong, the result can often be a blank page. In the case of viewports, even if our code is working, it may not be obvious because the viewports themselves are invisible. This section provides some tips and tools for debugging **grid** graphics code.

The `grid.ls()` function (see Section 7.2) can be used to list all grobs and viewports on the current page, although in some situations it will be useful to call `grid.force()` first (see Section 7.7).

One of the simplest things we can do to check where a viewport is placed on the page is to draw a rectangle around the current viewport with a simple call to `grid.rect()`.

When we are working with **grid** layouts, the `grid.show.layout()` function is useful for drawing a diagram to show how a page will be divided by a layout.

Finally, the **gridDebug** package provides the `gridTree()` function for drawing node-and-edge graphs to show the hierarchy of grobs and viewports in the current page. The package also includes a `grobBrowser()` function that generates a SVG version of the current **grid** output with tooltips for interactively exploring the names of grobs within a plot.

Chapter summary

It is possible to write simple **grid** graphics functions for the purpose of producing graphical output. Such functions should not assume that they have the entire device to draw into. They should only assume that they are drawing within a **grid** viewport. Naming any viewports created in the function and using `upViewport()` rather than `popViewport()` makes it possible for others to annotate the graphical output produced by the function. Naming all grobs produced by the function makes it possible for others to edit the output from the function (or remove grobs or add grobs or extract grobs).

Creating a graphical object to represent the output generated by the function requires extra effort to set up methods for the new graphical object class, but provides additional benefits. Most graphical objects will be gTrees consisting of a high-level description plus several child grobs representing the output produced. A gTree makes it possible for others to interact with the high-level description, while still being able to access the low-level element grobs. A grob can also be useful to provide information about the amount of space required to produce graphical output. Finally, a grob makes it possible for others to create higher-level gTrees with the grob as a child element.

Part III

THE GRAPHICS ENGINE

9

Graphics Formats

Chapter preview

This chapter describes how to produce graphical output in different formats. The output of graphics functions is typically drawn on screen initially, but this chapter describes how to save plots to files on disk. There is a discussion of the advantages and disadvantages of the various formats for different purposes. The same R code will sometimes produce slightly different output on different formats, so these differences are also described.

This part of the book is devoted to the core graphics engine in R, which is provided by the **grDevices** package. The information in this chapter and the next applies to almost all graphics functions and packages mentioned in this book.

The **grDevices** package is part of the standard R installation and is normally loaded by default in every R session. In a non-standard installation, it may be necessary to make the following call in order to access core graphics functions (if the **grDevices** package is already loaded, this will not do any harm).

```
> library(grDevices)
```

The graphics engine provides two main facilities for almost all graphics functions in R: support for producing output in different graphics formats, which is described in this chapter, and support for specifying values for graphical parameters, such as colors and fonts, which is described in Chapter 10.

9.1   Graphics devices

Throughout Parts I and II of this book, there have been vague statements about "the graphics window" or graphical output being drawn on a "page" or a "screen." This chapter addresses the issue of where graphical output appears and how it gets recorded.

In a typical interactive R session, a graphics window is automatically opened the first time that a graphics function is called and a plot is drawn on screen in this window. So for simple usage, there is no need for the user to decide where graphics output should go because there is a sensible default. However, for the purposes of producing a report, for example, in a PDF document, drawing a plot on screen is not very helpful. Instead, the plot needs to be saved in a PDF format, in a file on a hard disk. This section describes how to direct graphical output to a file rather than to the screen and how to specify the format of that file.

When using the popular R Studio* interface for R, there is a graphics "pane" where plots will appear. This is fine for casual use, but does not behave exactly like a normal R graphics device, so for achieving fine control over a plot it is advisable to explicitly open a graphics window (or file), as described in this chapter.

In R's terminology, graphical output is directed to a particular *graphics device*. In general, a graphics device must first be *opened*, then any subsequent calls to graphics functions produce output on that device. The `dev.new()` function opens the default device, as given by `options("device")`, but each device also has its own specific function. For example, the `pdf()` function opens a file and stores graphics output in a PDF format. A full list is given in the next section. For file-based devices, it is also important to *close* the device using the `dev.off()` function once all graphical output is complete.

The following code shows how to produce a simple scatterplot in PDF format. The output is stored in a file called `myplot.pdf`.

```
> pdf(file="myplot.pdf")
> plot(pressure)
> dev.off()
```

*https://www.rstudio.com/

A simple modification of this pattern produces the same output in PNG format (in a file called `myplot.png`), as shown below.

```
> png(file="myplot.png")
> plot(pressure)
> dev.off()
```

It is possible to have more than one device open at the same time, but only one device is currently *active* and all graphics output is sent to that device.

If multiple devices are open, there are functions to control which device is active. The list of open devices can be obtained using `dev.list()`. This gives the name (the device format) and number for each open device. The function `dev.cur()` returns this information only for the currently active device. The `dev.set()` function can be used to make a device active, by specifying the appropriate device number and the functions `dev.next()` and `dev.prev()` can be used to make the next/previous device on the device list the active device.

The `dev.size()` function can be used to obtain the size of the current device, in either inches, centimeters, or pixels.

All open devices can be closed at once using the function `graphics.off()`. When an R session ends, all open devices are closed automatically.

9.2 Graphical output formats

Table 9.1 gives a full list of functions that open devices and the output formats that they correspond to.

All of these functions provide several arguments to allow the user to specify things such as the physical size of the window or document that is created.

Due to differences between graphics formats, it is very unlikely that the same R code will produce *identical* results on different devices. For example, a PDF version of a plot is unlikely to appear identical to a PNG version of the same plot. Fonts are a particularly difficult feature to reproduce exactly across different formats.

Some of the distinct features of the various graphics formats are discussed further in the following sections.

Table 9.1

Graphics formats that R supports and the functions that open an appropriate graphics device.

Function	Graphical Format
Screen Devices	
x11() or X11()	X Window window (Cairo graphics)
windows()	Microsoft Windows window
quartz()	MacOS X Quartz window
File Devices	
postscript()	Adobe PostScript file
pdf()	Adobe PDF file
svg()	SVG file
win.metafile()	Windows Metafile file (Windows only)
png()	PNG file
jpeg()	JPEG file
tiff()	TIFF file
bmp()	BMP file
pictex()	LaTeX PicTeX file
xfig()	xfig FIG file
bitmap()	Multiple raster formats via Ghostscript

9.2.1 Vector formats

Graphics devices can be divided into two main groups: *vector* formats and *raster* formats. In a vector format, an image is described by a set of mathematical shapes, for example, a line segment from one (x, y) location to another. In a raster format, an image consists of an array of *pixels*, with information such as color recorded for each pixel. The vector-format version for drawing a line segment might look something like the following, which involves just the end points of the line that should be drawn.

```
2 2 moveto
8 6 lineto
```

By contrast, a raster format version of the same line might look like the following, which involves specifying which pixels should be drawn to show the line (the pixels with value 1 would be drawn).

```
0 0 0 0 0 0 0 0 0 0
0 0 0 0 0 0 0 0 1 0
0 0 0 0 0 0 1 1 0 0
0 0 0 0 1 1 0 0 0 0
0 0 1 1 0 0 0 0 0 0
0 1 0 0 0 0 0 0 0 0
0 0 0 0 0 0 0 0 0 0
```

Vector formats include PDF, PostScript, and SVG. Examples of raster formats are PNG, JPEG, TIFF, and all screen devices.

The R graphics engine is fundamentally vector based, so R plots are produced very faithfully on vector-based devices. When producing output on a raster device, the quality of the result may be lower than for a vector device, but this can be ameliorated by using a higher resolution (more pixels) or by using a raster device that implements *anti-aliasing*, which helps to produce smoother lines.

In general, vector formats are superior for images that need to be viewed at a variety of scales, but raster formats will produce much smaller files if the image is very complex. For most purposes, a vector format is usually the best choice, but it is sometimes more sensible to use a raster format when a plot is visually complex, for example, if it involves a large number of data points.

It may sometimes be necessary to make further modifications to an R plot using third-party software. In such cases, another consideration is that certain modifications of an image, for example removing a particular shape, are only possible with a vector format. On the other hand, other modifications, such as making all white pixels in an image transparent, are easier with raster formats. Because it is easy to convert a vector format to a raster version, while the reverse is very difficult if not impossible, it usually makes sense to produce a vector image from R if the image will be modified later.

PDF

PDF is a good choice of format, partly because of the widespread availability of viewing software such as Adobe Reader. It is also a very sophisticated format, so it is able to faithfully produce anything that R graphics can do.

The primary device for producing R plots in PDF format is the `pdf()` device.

The first argument is the name of the file to produce. By default, this will produce a single file, which can contain several pages of output. Section 9.5 describes how to produce a separate file for each page of output.

By default the `pdf()` device produces a seven-inch square document, but a custom physical size can be specified as the `width` and `height`, in inches. It is also possible to specify a standard paper size, e.g., `"a4"`, via the `paper` argument. However, this paper size is independent of the width and height of the actual plot *unless* the width and height are set to zero, in which case the plot expands to fit the paper size (minus 0.25 inch margins).

The default (sans-serif) font for the `pdf()` device is Helvetica, but a different default can be specified via the `family` argument. For example, `"serif"` uses a Times font, and `"mono"` produces Courier. Much more information about selecting fonts is provided in Section 10.4.

In non-English locales, it may be necessary to specify an appropriate `encoding` for the file, although the `pdf()` function makes some attempt to automate this.

R also provides some support for locales with very large character sets, such as Chinese *hanzi*, Japanese *kanji*, and Korean *hanja*. For these cases, there are several predefined *CID-keyed* fonts, which are also included in the list produced by `pdfFonts()`. It is also possible to define new fonts via the `CIDFont()` function, but this does require detailed knowledge of the relevant font technology.

The `pdf()` device does *not* embed fonts within the PDF file. This is significant because PDF viewer software will *substitute* fonts if they are not embedded within a PDF file and they are not available on the system where the file is being viewed. If a non-standard font is used and font substitution occurs, the resulting plot may have missing characters or at best look quite untidy. This means that a plot should only use fonts that are known to be installed on the system where the plot is to be viewed (e.g., the default Helvetica, Times, or

Courier fonts), *or* all fonts should be embedded within the PDF file using the embedFonts() function. In the latter case, all relevant fonts must be installed on the system that is used to perform the embedding.

In summary, any plot that makes use of the standard fonts should be fine, but any plot that makes use of more exotic fonts should call embedFonts() to make sure that the plot can be viewed or printed properly on any system.

When saving graphics that include text in a PDF format, the default behavior is to use *kerning* to make small adjustments to the positioning of certain pairs of characters. For example, a lowercase 'a' beside an uppercase 'T' are placed closer together than a lowercase 'a' beside a lowercase 'o'. This facility is turned on or off via the useKerning argument.

Another special situation arises when drawing polygons that self-intersect. There are two main algorithms for determining the interior of such polygons: the *non-zero winding rule* and the *even-odd rule*. Unfortunately, the R graphics engine does not explicitly specify a *fill rule* for self-intersecting polygons, so the default is to use the non-zero winding rule. The fillOddEven argument can be used to change to the even-odd rule instead.

Another way to produce PDF output in R is to use the function that is based on the Cairo graphics library,* cairo_pdf(). The advantage of this function is that it may provide better support for fonts, including automatic embedding of fonts, although this does depend on the installation of further software libraries. The downside to this PDF device is that it will sometimes generate raster output for complex images.

PostScript

PostScript can be thought of as a predecessor of PDF. In some ways, PostScript is actually more sophisticated than PDF, but it does not support some of the more modern features such as semitransparent colors and hyperlinking. This means that PostScript output cannot faithfully produce everything that R graphics can do.

The main way to produce PostScript output is using the postscript() device. This shares many features with the pdf() device as described above, including the ability to size the device, text kerning, and polygon fill rules.

Device sizing is slightly different in that the paper setting is dominant over the width and height. For example, on an "a4" PostScript page, the plot will fill the page by default. The PostScript produced by R is compatible with *Encapsulated* PostScript (EPS), which is useful for including R plots

---

*http://cairographics.org/.

within other documents (see Section 9.3), but to control the size of a plot it is necessary to specify `paper="special"` *as well as* an appropriate width and height. In this situation, it is usually also a good idea to specify a *portrait* orientation for the page via `horizontal=FALSE`. The `setEPS()` function is useful for setting up appropriate default settings for Encapsulated PostScript output.

Another difference between the PostScript device and the PDF device is that *all* fonts that are used in a PostScript plot must be "predeclared" via the `fonts` argument when the device is first opened.

One limitation with the `postscript()` device is that it does not support semitransparency. Any attempt to draw a semitransparent color will fail with a warning. If PostScript is the required format, one avenue is to produce PDF and then convert to PostScript using third-party software such as ImageMagick.* Another option is to use the Cairo-based device `cairo_ps()`. However, both of those options are likely to produce raster elements within the PostScript file, which means that the quality of the image may be reduced.

SVG

SVG is a format with tremendous potential because it offers an open standard vector format, as sophisticated as the PDF format, that can be embedded in web pages. All modern web browsers now support SVG. SVG output can be produced with the `svg()` device.

Because of the limitations of the R graphics engine, it is not possible to take advantage of more advanced SVG features, such as compositing operators and animation through the `svg()` device. However, Chapter 13 describes the **gridSVG** package, which provides access to a wide range of advanced SVG features. Some of the extension packages described in Section 9.7 also provide access to some extra SVG features.

Windows Metafile

The Windows Metafile format is important because it is the vector format that should be most compatible with Microsoft products such as Word, Excel, and PowerPoint. This format can only be produced on Windows systems. A Windows Metafile file can be generated with the `win.metafile()` function.

---

*`http://www.imagemagick.org/`.

9.2.2   Raster formats

The raster device that users will encounter most often is the graphics window on screen. This is the quickest and simplest way to view graphical output. Screen devices are different on different operating systems: typically, a Cairo-based X Window device on Linux, a Quartz device on MacOS X, and a native Windows device on Windows. There are some differences between these devices (see Section 9.4), so R code is unlikely to produce identical results on different platforms. On Linux and MacOS X there is also an X Window device, which lacks support for some graphics features, but is faster than the Cairo-based device.

When saving graphics to a file, there are several raster formats to choose from. The PNG format is desirable because it is *lossless*, which means that it compresses the image (most raster formats compress the image to save space) in such a way that no information is lost. This means that a PNG file can be edited without reducing the quality. The JPEG format, by comparison, uses *lossy* compression so, although JPEG files will typically be smaller than PNG files, repeatedly editing a JPEG will result in a reduction in quality. Furthermore, the JPEG compression is better suited to complex images with lots of different regions (like photographs), whereas the PNG format does a better job with simpler images that include lines and text and large areas of constant color. Consequently, the PNG format is usually better for statistical plots, though an exception might be a very busy `image()` plot or `contour()` plot.

The JPEG format does not support semitransparency. The PNG format does, but this is only partially supported on Windows, and only via the default Cairo-based devices on Linux and MacOS X.

Neither PNG nor JPEG formats support multiple pages in a document, so if a `png()` device is opened and then more than one page of output is produced, the result will be several PNG files rather than just one (by default, the file names are automatically numbered).

TIFF is a very sophisticated format that allows multiple pages of raster output within a single file. It is less well supported by web browsers, but may be the preferred format for publishers of books or journal articles.

Determining the size of a raster image is less straightforward than it is for vector formats. The `width` and `height` of a raster device are specified as a number of pixels rather than as a physical size in inches. The physical size of a raster image is then determined by the *resolution* at which it is viewed. For example, a PNG image that is 72 pixels wide will be 1 inch wide when viewed on a screen with a resolution of 72 dpi (dots per inch), but it will be only 0.75 inches wide on a screen with a resolution of 96 dpi.

It is possible to specify a fixed resolution for a raster format image via the `res` argument. However, this information will not necessarily be respected when the image is displayed. For example, a web browser may just use the resolution of the screen when displaying images on web pages (so the image size will still vary depending on the screen resolution). On the other hand, if a raster image is included within a LaTeX document, the resolution of the image is respected.

As a general rule of thumb, if a raster image is being prepared for use on a web page, there is no point in worrying about setting the resolution, but if a raster image is being prepared for inclusion in a document that is to be typeset, such as a LaTeX or Microsoft Word document, then setting the resolution may be worthwhile, particularly if a high-quality image is required.

Because the physical size of a raster image can be ambiguous, it can be difficult to control the size of text in a raster image. The `pointsize` argument specifies the default size of text for an image, but what this means is again dependent on the resolution at which the image is displayed. The size of text is given in *big points* ($\frac{1}{72}$ inch), *relative to the* `res` *argument*. This means that the size of text is calculated as if the resolution of the image is going to be respected. The result should be as expected when a raster image is included in another document, but the result can be confusing if the image is displayed at screen resolution.

In summary, the physical size of text in a plot depends on the size of the text, the size of the image, the resolution of the image, and whether the image is displayed at screen resolution or at the native resolution of the image.

9.2.3   R Studio

The R Studio IDE for R includes a "plot pane" where graphics output will appear by default. This plot pane is not a standard R graphics device. It is useful for exploratory graphical analysis, but if you want to control the exact appearance of the final result, you should open and use an explicit R graphics device, for example, by calling `pdf()` or `png()`.

This warning applies generally for all "GUI" R interfaces; in order to obtain the best results for graphical output, rather than using a "Save as" menu option, the recommended approach is to explicitly control the opening and closing of R graphics devices.

9.3 Including **R** graphics in other documents

There are two typical uses of R graphics. One is to produce basic plots on screen for exploratory data analysis, and the other is to produce finely tuned plots in a file format for inclusion in a larger document such as a web page or a printed report. This section deals with some issues specifically related to the latter task.

One important issue to consider is the physical size of text and the physical width of lines in a plot within the final document. Text has to be readable and lines typically need to be wide enough for print resolution so that, for example, they do not disappear when photocopied.

The default, for vector formats, is to produce a seven-inch square document, using a 12-point, sans-serif font, with lines $\frac{1}{96}$ inches wide. This is fine for viewing a plot on its own, but is much too large for a typical document, for example, when including a plot in a figure within an A4 page.

The best approach is to produce the plot at the size that it needs to be in the final document and specify the appropriate font size and line width explicitly.

9.3.1 LaTeX

Standard vector formats such as PDF and PostScript are ideal for including within LaTeX documents. However, there is one situation where a more LaTeX-specific option may be more desirable.

One thing that LaTeX does exceptionally well is the typesetting of mathematical formulae. R's mathematical annotation facility attempts to emulate LaTeX, but it is not as good as the real thing, particularly when the fonts involved are not the TeX math fonts.

There is a special **cmsyase** font that can be used to draw mathematical formulae in R with TeX math fonts. This is available from the **fontcm** extension package for R.

One way to produce graphics output specifically for inclusion in a LaTeX document is to use the **pictex()** device. This produces LaTeX macros from the PiCTeX package to draw a plot. The main advantage of this is that the text in the plot will use the same font as the rest of the LaTeX document. Unfortunately, this device is very rudimentary, so it is not suitable for anything other than very basic plots (it does not even support colors). See Section 9.7 for a more sophisticated alternative.

9.3.2 "Productivity" software

Microsoft software products have a tendency to play nicely with each other and with Microsoft formats, but less well with other software products and formats. This is particularly true for vector graphics formats, so possibly the best vector format for including plots in Microsoft products, such as Word and Excel, is the WMF format (Windows Meta-File). Microsoft products should cope well with the standard raster formats, though there is also the Windows-specific BMP format.

The Open Office software has better support for including PDF plots in documents and will also cope with standard raster formats.

9.3.3 Web pages

Historically, the standard way to include an image in a web page has been to use a raster format, such as PNG. However, due to the improved support in modern web browsers, SVG is now the preferred format.

9.4 Device-specific features

Not all graphics devices are created equal. The same R code can produce slightly different graphical output depending on the graphics device format.

While the performance of vector devices should be quite consistent on all platforms (Windows, Linux, MacOS X), the performance of raster devices is much more platform dependent. On the other hand, for a specific platform, plots saved in a raster format should have the same appearance as they do on-screen.

One area where differences can become evident is in the selection of fonts. The standard set of fonts, as described in Section 10.4, should always be available, though there will be small differences in appearance on different platforms (e.g., the default "sans" font is Arial on Windows and Helvetica on Linux). Section 10.4 provides details on how to select fonts for different graphics devices.

On some devices, the font size that is specified will not be honored exactly. For example, when drawing in a raw X Window window with bitmap fonts, there are only a finite set of font sizes available and this set will vary depending on which fonts are installed. For the PostScript and PDF formats, font sizes should scale appropriately to any size.

Anti-aliasing can dramatically improve the quality of a raster image by smoothing the appearance of lines and text. The support for anti-aliasing will vary across graphics devices. If the purpose is to include a raster image in another document, then generating a high-resolution image is another way to improve quality.

The Windows screen device has less-complete support for semitransparent colors, compared to the default screen device on Linux and MacOS X.

On Linux and MacOS X, where the default screen and raster devices are Cairo based, it is also possible to produce screen output and raster formats directly via the X Window system. This typically produces a poorer quality image, for example, there is no support for semitransparent colors or anti-aliasing, but the rendering is faster so this option could be considered for particularly complex images.

An alternative way to produce raster format images that should produce more consistent results across platforms is to use the `bitmap()` function. The downside is that this requires the installation of additional software (Ghostscript). Section 9.7 describes some other possibilities for producing consistency across platforms.

9.5 Multiple pages of output

For a screen device, starting a new page involves clearing the window before producing more output. Some "GUI" interfaces for R provide a "plot history" facility for revisiting previous screens of output, but on most raw on-screen devices, the output of previous pages is lost.

If a piece of code produces several pages of plots, the `devAskNewPage()` function can be used to force a user prompt before each new page is started. This allows the user to view each page at leisure before indicating to R to move on to the next page.

For file devices, the output format dictates whether multiple pages are supported. For example, PostScript and PDF allow multiple pages, but PNG does not. It is usually possible, especially for devices that do not support multiple pages of output, to specify that each page of output produces a separate file. This is achieved by specifying the argument `onefile=FALSE` when opening a device and specifying a pattern for the file name like `file="myplot%03d"` so that the `%03d` is replaced by a three-digit number (padded with zeroes) indicating the "page number" for each file that is created.

9.6 Display lists

R maintains a *display list* for each open device, which is a record of the output on the current page of a device. This is used to redraw the output when a device is resized and can also be used to copy output from one device to another.

The function `dev.copy()` copies all output from the active device to another device. The copy may be distorted if the aspect ratio of the destination device — the ratio of the physical height and width of the device — is not the same as the aspect ratio of the active device. The function `dev.copy2eps()` is similar to `dev.copy()`, but it preserves the aspect ratio of the copy and creates a file in EPS (Encapsulated PostScript) format that is ideal for embedding in other documents (e.g., a LaTeX document). The `dev2bitmap()` function is similar in that it also tries to preserve the aspect ratio of the image, but it produces one of the output formats available via the `bitmap()` device.

The function `dev.print()` attempts to print the output on the active device. By default, this involves making a PostScript copy and then invoking the print command given by `options("printcmd")`.

The display list can consume a reasonable amount of memory if a plot is particularly complex or if there are very many devices open at the same time. For this reason, it is possible to disable the display list by typing the expression `dev.control(displaylist="inhibit")`. If the display list is disabled, output will not be redrawn when a device is resized, and output cannot be copied between devices.

There is also a `recordPlot()` function, which saves the display list to an R variable. The variable can then be passed to the `replayPlot()` function to draw the saved plot.

Table 9.2

Graphics formats that are provided by extension packages for R and the functions that open an appropriate graphics device.

Function	Graphical Format	Package
Cairo()	Multiple formats	**Cairo**
tikz()	LaTeX PGF/TikZ file	**tikzDevice**
devSVGTips()	SVG file	**RSVGTipsDevice**

9.7   Extension packages

Several extension packages for R provide a number of extra graphical formats that are not provided by the **grDevices** package itself. In general, these work just like the core devices, with a function provided to open a device in the appropriate format. Additional functions may be provided for handling other features of the device, such as fonts. Table 9.2 lists some of the extension packages that provide graphics devices.

The usefulness of the **Cairo** package is that it allows Cairo-based graphics output on any platform (although it requires the Cairo graphics library to be installed first). This has the advantage that the output on a Cairo-based screen device should be very similar on all platforms and the output on a Cairo-based file device should be very similar to the output on a screen device.

The **tikzDevice** package provides a sophisticated solution for producing graphical output for inclusion in LaTeX documents. The main advantages are that the fonts for text in the plot will match the fonts used in the LaTeX document and LaTeX's native mathematical formula syntax can be used for text in plots.

The **RSVGTipsDevice** package provides an alternative way to produce SVG output, with the advantage of allowing tooltips and hyperlinks to be added to the SVG file. See Chapter 13 for the **gridSVG** package, which takes this idea much further.

Chapter summary

R graphics can produce a wide variety of graphical formats. In interactive use, graphics output is drawn on screen, but it is also possible
to save graphics output in a file. A vector graphics format usually produces a better-quality result than a raster format when saving plots
to a file, but the choice of format will also depend on how the plot
will be used (e.g., included in a LaTeX document versus distributed as
part of a web page).

10

Graphical Parameters

Chapter preview

This chapter describes how to specify graphical parameters, including information about specifying colors, how to generate sets of coherent colors, information about how to specify fonts for drawing text, and information about how to produce special symbols and formatting for drawing mathematical formulae. The information in this chapter is useful for controlling the output of almost all graphics functions in R.

Graphical parameters are the arguments to functions that influence the detailed appearance of a graphical image. They apply the make-up to the basic bone structure of an image. Examples include the color and line width used to draw a line and the font used to draw text.

Despite the fact that the R graphics universe consists of two distinct graphics systems, base and **grid**, the way that graphical parameters are specified is quite consistent across both of these systems.

10.1   Colors

The easiest way to specify a color in R is simply to use the color's name. For example, `"red"` can be used to specify that graphical output should be (a very bright) red. R understands a fairly large set of color names; type `colors()`

(or `colours()`) to see a full list of known names.

It is also possible to specify colors using one of the standard color space descriptions. For example, the `rgb()` function allows a color to be specified as a Red-Green-Blue (RGB) triplet of intensities. Using this function, the color red is specified as `rgb(1, 0, 0)`. The function `col2rgb()` can be used to see the RGB values for a particular color name (although the resulting color channels are in the range 0 to 255 rather than 0 to 1).

```
> col2rgb("red")


      [,1]
red    255
green    0
blue     0
```

An alternative way to provide an RGB color specification is to provide a string of the form `"#RRGGBB"`, where each of the pairs `RR`, `GG`, `BB` consist of two hexadecimal digits giving a value in the range zero (`00`) to 255 (`FF`). In this specification, the color red is given as `"#FF0000"`.

In R, RGB color specifications are interpreted relative to the sRGB color space (IEC standard 61966).*

There is also an `hsv()` function for specifying a color as a Hue-Saturation-Value (HSV) triplet. The terminology of color spaces is fraught, but roughly speaking: *hue* corresponds to a position on the rainbow, from red (0), through orange, yellow, green, blue, indigo, to violet (1); *saturation* determines whether the color is dull (grayish) or bright (colorful); and *value* determines whether the color is light or dark. The HSV specification for the (very bright) color red is `hsv(0, 1, 1)`. The function `rgb2hsv()` converts a color specification from RGB to HSV.

```
> rgb2hsv(255, 0, 0)


   [,1]
h    0
s    1
v    1
```

A better alternative to either `rgb()` or `hsv()` is the `hcl()` function. Similar to `hsv()`, this function specifies colors as a hue, a *chroma* (or colorfulness,

---

*http://www.color.org/chardata/rgb/srgb.xalter.

similar to saturation), and a *luminance* (or lightness, similar to value). The color "red" corresponds to hcl(12, 179, 53).

The hcl() function is better than the hsv() function because it works in the (polar) CIE-LUV color space, in which a unit distance is close to a perceptually constant change in color, so, for example, holding chroma and luminance constant while varying only hue produces colors that are approximately similar in their visual impact on the observer.

Greyscale colors can be generated using the function grey() (or gray()). These functions take a vector of numeric values between 0 (black) and 1 (white).

One final way to specify a color is simply as an integer index into a predefined set of colors. The predefined set of colors can be viewed and modified using the palette() function. In the default palette, red is specified as the integer 2.

10.1.1   Semitransparent colors

All R colors are stored with an alpha transparency channel. An alpha value of 0 means fully transparent and an alpha value of 1 means fully opaque. When an alpha value is not specified, the color is opaque.

The function rgb() can be used to specify a color with an alpha transparency channel, simply by providing a fourth value to the function. For example, rgb(1, 0, 0, 0.5) specifies a semitransparent red. Alternatively, a color can be specified as a string beginning with a "#" and followed by *eight* hexadecimal digits. In that case, the last two hexadecimal digits specify an alpha value in the range 0 to 255. For example, "#FF000080" specifies a semitransparent red.

A color may also be specified as NA, which is usually interpreted as fully transparent (i.e., nothing is drawn). The special color name "transparent" can also be used to specify full transparency.

WARNING: If a graphic device does not support semitransparency, semi-transparent colors are rendered as fully transparent.

10.1.2   Converting colors

There are many other ways to specify colors besides the RGB, HSV, and polar CIE-LUV color spaces described so far and the `convertColor()` function provides a mechanism for converting between different color spaces.

The following code shows an example where the color `"red"` is converted to the CIE-LUV color space. This can be a useful transformation because the L component of the result can be used to convert the color to grayscale. The `col2rgb()` function is used to obtain a matrix containing the separate red, green, and blue components, those are normalized to a zero-to-one range by dividing by 255, and then the matrix is transposed so that the components are different columns. The transformation is from R's native color space, sRGB, to CIE-LUV.

```
> convertColor(t(col2rgb("red")/255), "sRGB", "Luv")
```

```
            L        u        v
[1,]  53.48418  175.3647  37.80017
```

The L component of the result corresponds to the values given for the `hcl()` specification of `"red"` on page 321. The u and v components do not correspond to the h and c components of the `hcl()` example because the `hcl()` function works in polar coordinates, whereas u and v are cartesian dimensions within the CIE-LUV color space.

Another useful tool is the `adjustcolor()` function, which allows the components of an existing color to be scaled. For example, the following code takes the color `"red"` and makes it semitransparent.

```
> adjustcolor("red", alpha.f=.5)
```

```
[1]  "#FF000080"
```

This result corresponds to the explicit color specification for semitransparent red that was given above.

The **colorspace** package provides more tools for converting between a wider range of color spaces.

10.1.3   Color sets

More than one color is often required within a single plot, for example to distinguish between different groups of data symbols, and in such cases it

Table 10.1
Functions to generate color sets. R functions that can be used to generate coherent sets of colors.

Name	Description
`rainbow()`	Colors vary from red through orange, yellow, green, blue, and indigo, to violet.
`heat.colors()`	Colors vary from white, through orange, to red.
`terrain.colors()`	Colors vary from white, through brown, to green.
`topo.colors()`	Colors vary from white, through brown then green, to blue.
`cm.colors()`	Colors vary from light blue, through white, to light magenta.
`gray.colors()`	A set of shades of gray.

can be difficult to select colors that are aesthetically pleasing or are related in some way (e.g., a set of colors in which the brightness of the colors decreases in regular steps). Table 10.1 lists some functions that R provides for generating sets of colors. Each of these functions takes a single numeric argument and returns that number of colors. For example, the following code produces five colors from the `rainbow()` function.

```
> rainbow(5)
```

```
[1] "#FF0000FF" "#CCFF00FF" "#00FF66FF" "#0066FFFF"
[5] "#CC00FFFF"
```

The output of the expression `example(rainbow)` provides a nice visual summary of the color sets generated by several of these functions.

Each of the functions in Table 10.1 (apart from `gray.colors()`) selects a set of colors by taking regular steps along a path through the HSV color space. As mentioned previously, a more perceptually uniform set of colors can be obtained by working in the CIE-LUV color space. For example, the following code generates six colors from the CIE-LUV color space that vary regularly in terms of hue, but are all equally bright (the chroma component is fixed at 50) and all equally light (the luminance component is fixed at 60).

```
> hcl(seq(0, 300, 60), 50, 60)
```

```
[1] "#C87A8A" "#AC8C4E" "#6B9D59" "#00A396" "#5F96C2"
[6] "#B37EBE"
```

There are a number of extension packages that provide further functions for generating a set of colors, for example, **RColorBrewer** and **pals**.

The functions `colorRamp()` and `colorRampPalette()` are a little different because they are not color set generators. Instead, they are color set *function* generators. These functions accept a set of colors and color space to work in and they interpolate a path through the color space (either joining the starting colors with straight lines or interpolating a smooth curve through the colors), then they return a function that can be called to select colors from the interpolated path.

One difference between the functions is that `colorRamp()` produces a function that can generate colors based on a sequence of values in the range 0 to 1, like `gray.colors()`, whereas `colorRampPalette()` produces a function that can generate *n* colors, like `rainbow()`.

Another difference between the functions is that `colorRamp()` returns a matrix of red, green, and blue color components, whereas `colorRampPalette()` returns a vector of colors.

The following code demonstrates `colorRampPalette()` being used to create a color set generating function that produces colors ranging from `"blue"` to `"gray"`. The function is then used to generate five colors.

```
> bluegray <- colorRampPalette(c("blue", "gray"))
> bluegray(5)
```

```
[1] "#0000FF" "#2F2FEE" "#5F5FDE" "#8E8ECE" "#BEBEBE"
```

10.1.4   Device dependency of color specifications

The colors that R sends to a graphics device are sRGB colors. This should be appropriate for drawing to a screen device because most computer monitors are set up to work with sRGB. Also, colors used on web pages are typically sRGB, so raster file formats produced by R, such as PNG, should work reasonably well there too.

However, the final appearance of a color can vary considerably when it is viewed on a screen, or printed on paper, or displayed through a projector as it depends on the physical characteristics of the screen, printer ink, or projector. When an image is saved in a PDF or PostScript format, R records the fact that sRGB colors are being used so printers and viewers have some chance of producing the right result.

10.2 Line styles

It is possible to control the width of a line, the pattern used to draw the line (e.g., solid versus dashed), and the styling used for the ends and corners of a line.

10.2.1 Line widths

The width of lines is specified by a simple numeric value, e.g., `lwd=3`. This value is a multiple of 1/96 inch, with a lower limit of 1 pixel on some screen devices. The default value is `1`.

10.2.2 Line types

R graphics supports a fixed set of predefined line types, which can be specified by name, such as `"solid"` or `"dashed"`, or as an integer index (see Figure 10.1). In addition, it is possible to specify customized line types via a string of digits. In this case, each digit is a hexadecimal value that indicates a number of "units" to draw either a line or a gap. Odd digits specify line lengths and even digits specify gap lengths. For example, a dotted line is specified by `lty="13"`, which means draw a line of length one unit then a gap of length three units. A unit corresponds to the current line width, so the result scales with line width, but is device dependent. Up to four such line-gap pairs can be specified. Figure 10.1 shows the available predefined line types and some examples of customized line types.

10.2.3 Line ends and joins

When drawing thick lines, it becomes important to select the style that is used to draw corners (joins) in the line and the style that is used to draw the ends of the line. R provides three styles for both cases: line endings can be

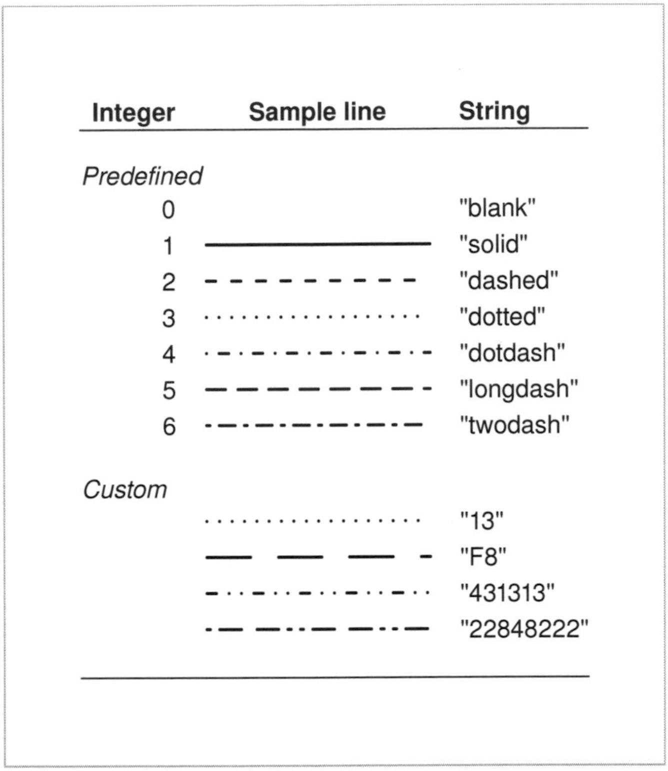

Figure 10.1
Predefined and custom line types. Line type may be specified as a predefined integer, as a predefined string name, or as a string of hexadecimal characters specifying a custom line type.

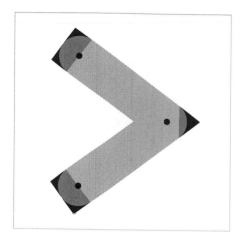

Figure 10.2
Line join and line ending styles. Three thick lines have been drawn through the same three points (indicated by black circles), but with different line end and line join styles. The black line was drawn first with "square" ends and "mitre" joins; the dark gray line was drawn on top of the black line with "round" ends and "round" joins; and the light gray line was drawn on top of that with "butt" ends and "bevel" joins.

"round" or flat (with two variations on flat, "square" or "butt"); and line joins can be "mitre" (pointy), "round", or "bevel". The differences are most easily demonstrated visually (see Figure 10.2).

When the line join style is "mitre", the join style will automatically be converted to "bevel" if the angle at the join is too small. This is to avoid excessively pointy joins. The point at which the automatic conversion occurs is controlled by a miter limit, which specifies the ratio of the length of the miter divided by the line width. The default value is 10, which means that the conversion occurs for joins where the angle is less than 11 degrees. Other standard values are 2, which means that conversion occurs at angles less than 60 degrees, and 1.414, which means that conversion occurs for angles less than 90 degrees. The minimum miter limit value is 1.

It is important to remember that line join styles influence the corners on rectangles and polygons as well as joins in lines.

10.3   Data symbols

The data symbol used for plotting points is specified as either an integer, which indexes one of 26 predefined data symbols (see Figure 10.3), or directly as a single character. Some of the predefined data symbols (`pch` between 21 and 25) allow a fill color separate from the border color.

Integer values larger than 32, but less than 127, are interpreted as ASCII character values and the corresponding character is drawn.

If `pch` is a character, then that letter is used as the plotting symbol. The character "." is treated as a special case and the device attempts to draw a very small dot.

The `text()` and `grid.text()` functions can also be used to draw a single character, which expands the range of possible "plotting symbols" enormously (depending on our font and system locale).

10.4   Fonts

Whenever we draw text as part of a plot or image, R needs to know what font to use for the text. Specifying a font in R consists of specifying a font *family*, such as Helvetica or Courier, and specifying a font *face*, such as **bold** or *italic*. If we specify nothing, the default should be a plain sans-serif font (e.g., Helvetica or Arial).

The following code provides a simple demonstration (using **grid** functions): the first expression draws text with the default (sans-serif) font; the second expression specifies that the font family should be serif; and the third expression specifies that the font face should be bold. The result is shown in Figure 10.4.

```
> grid.text("hello", x=1/4)
> grid.text("hello", x=2/4,
            gp=gpar(fontfamily="serif"))
> grid.text("hello", x=3/4,
            gp=gpar(fontfamily="serif", fontface="bold"))
```

The code above demonstrates that the font family and font face are specified

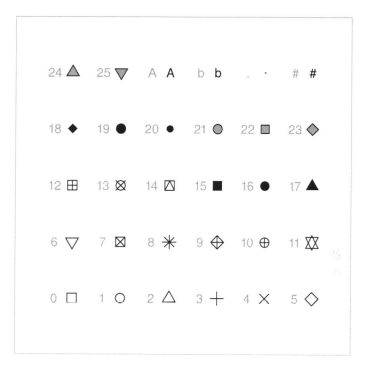

Figure 10.3
Data symbols available in R. A particular data symbol is selected by specifying an integer between 0 and 25 or a single character. In the diagram, the relevant integer or character value is shown in gray to the left of the relevant symbol.

Figure 10.4
Specifying fonts in R. The "hello" on the left is in the default plain sans-serif font. The middle "hello" has a serif font family specified and the right "hello" has, in addition, a bold font face specified.

Figure 10.5
Special characters from the Hershey font set.

just using character values, e.g., `"serif"` and `"bold"`. Section 10.4.1 describes the set of character values that R allows for specifying the font family and Section 10.4.2 describes the same for font face.

This simple method for specifying fonts in R does not allow us to select from the full range of fonts on a system; Section 10.4.1 describes more details about fonts in general and some ways to work around R's limitations. It is also only possible to specify a single font for each function call that draws text (e.g., it is not possible to *emphasize* a single word within a string of text); Section 10.5 offers a partial workaround for this limitation.

10.4.1   Font family

This section describes the values that can be used to specify the font family.

There is a set of *device-independent fonts* that are supported on all graphics devices. These are `"sans"`, which gives a sans-serif font, like Arial; `"serif"`, which gives a serif font, like Times; and `"mono"`, which gives a monospace font, like `Courier` (see Table 10.2). The default font family for most graphics devices is `"sans"`.

In addition, the Hershey outline fonts are also distributed with R and are available for *all* output formats. The full set of Hershey font family names are shown in Table 10.2. These fonts are *outline fonts* and are rendered by simply tracing the outline of each character. To a large extent, the Hershey fonts are a low-quality anachronistic curiosity, but they are possibly the simplest way to get exactly the same text result on all possible output devices and platforms. The Hershey fonts are also a source of some unusual characters and idiograms; for example, the following code was used to draw Figure 10.5. The output from `demo(Hershey)` includes a full set of these special characters.

```
> chars <- sprintf("\\#H%04d", 861:866)
> chars
> grid.text(chars, x=1:6/7, gp=gpar(fontfamily="HersheySans"))
```

Beyond these device-independent font family specifications, it is also possible to use specific font family names, such as `"Comic Sans"`. However, the range

Table 10.2

Device-independent and Hershey font families that are distributed with R. A font family is specified as a character value.

Name	Description
Device-independent fonts	
`"serif"`	Serif variable-width font
`"sans"`	Sans-serif variable-width font
`"mono"`	Mono-spaced "typewriter" font
Hershey fonts	
`"HersheySerif"`	Serif variable-width font
`"HersheySans"`	Sans-serif variable-width font
`"HersheyScript"`	Serif "handwriting" font
`"HersheyGothicEnglish"`	Gothic script font
`"HersheyGothicGerman"`	Gothic script font
`"HersheyGothicItalian"`	Gothic script font
`"HersheySymbol"`	Serif symbol font
`"HersheySansSymbol"`	Sans-serif symbol font

of valid font names and the amount of setup required depends on two things: first, we must have the font installed on our system (it is also important to consider the audience for the plot and whether they will have the font installed); and, second, the specification of a font is different depending on the graphics device that we are drawing on. The following sections cover the various options for different standard graphics devices.

In brief, if you want to use a specific font, the easiest approach will be to use one of the Cairo-based graphics devices: the default screen device on Linux, `cairo_pdf()` for PDF output, or the **Cairo** package on Windows. The best last resort, if nothing else is working, is to try the **showtext** package. Read on for a much more detailed discussion.

Screen fonts

When we draw text to an on-screen graphics device, the fonts that are used will usually depend on our operating system.

On Windows, the screen device can make use of any font that is installed on the computer, but the font must be "registered" with R first using the `windowsFont()` and `windowsFonts()` functions. For example, if we know that the font `Algerian` is installed on our computer, the following code registers that font with R under the name `"AG"` and then draws text with that font by specifying the font family "AG".

```
> windowsFonts(AG=windowsFont("Algerian"))
> grid.text("hello", gp=gpar(fontfamily="AG"))
```

The **extrafont** package provides several useful functions that make this process easier: the font_import() function can be used to discover the complete list of fonts that are installed on our computer; and the loadfonts() function can be used to register all of the installed fonts with R at once.

```
> library(extrafont)
> font_import()
> loadfonts(device="win")
```

There is also a fonts() function that lists the fonts that have been discovered; the code and output below shows the first few fonts that were discovered on a Windows 10 computer.

```
> head(fonts())
```

```
[1] "Agency FB"  "Algerian"   "AR BERKLEY" "AR BLANCA"
[5] "AR BONNIE"  "AR CARTER"
```

After also calling loadfonts(), this means that I can use the following code to draw text with the "AR BONNIE" font.

```
> grid.text("hello", gp=gpar(fontfamily="AR BONNIE"))
```

On Linux, the default screen device is most likely to be a Cairo graphics device. In this case, we again have access to any font that is installed on the computer and selecting a specific font is even easier—we can simply specify the font family name (there is no "registration" step).

The **gdtools** package provides a sys_fonts() function to list all fonts that are available to a Cairo device. For example, the code and output below show the first few font names available on an Ubuntu 16.04 computer.

```
> library(gdtools)
> fonts <- sys_fonts()
> head(as.character(fonts$family))
```

```
[1] "Tlwg Mono"          "Courier New"        "Gillius ADF"
[4] "STIXIntegralsUpD" "STIXIntegralsD"      "NanumMyeongjo"
```

The following code shows that, on that Ubuntu computer, any of those font names can be used directly when drawing text.

```
> grid.text("hello", gp=gpar(fontfamily="Tlwg Mono"))
```

Another option on Linux is to use a plain X Window graphics device. A plain X Window graphics device does not produce the best quality output, so we are less likely to be worried about tweaking details like the font. Nevertheless, if we want to use a non-standard font, we must first define the font using the X11Font() function, and then register it with R using the X11Fonts() function.

The X11Fonts() function can be used both to view existing fonts and to define new ones. The code below provides an example of the former use, which allows us to see the format of a X Window font specification.

```
> X11Fonts("sans")
```

```
$sans
[1] "-*-helvetica-%s-%s-*-*-%d-*-*-*-*-*-*-*"
```

The following code shows how we can set up a new font. Having determined that an Ubuntu computer has a set of "bitstream charter" fonts installed (e.g., using the shell utility xlsfonts), we use the X11Font() function to create a new font description and then register that font with R using X11Fonts().

```
> charterFont <-
      X11Font("-*-bitstream charter-%s-%s-*-*-%d-*-*-*-*-*-*-*")
> X11Fonts(charter=charterFont)
```

We can now use "charter" as a font family specification when drawing text on an X Window device.

```
> grid.text("hello", gp=gpar(fontfamily="charter"))
```

On a MacOS X computer, the default screen device is a Quartz graphics device. As for Windows and X Window, there is a quartzFont() function for defining a new font family and a quartzFonts() function for registering the font family with R. Defining a MacOS X font requires four font names, one each for normal font face, bold, italic, and bold-italic. The following code shows an example that registers an "Avenir" font.

```
> avenirFont <- quartzFont(c("Avenir Book", "Avenir Black",
                             "Avenir Book Oblique",
                             "Avenir Black Oblique"))
> quartzFonts(avenir=avenirFont)
```

We can now use the "avenir" font to draw text on a Quartz device.

```
> grid.text("hello", gp=gpar(fontfamily="avenir"))
```

Fonts for raster formats

When we use a raster graphics device, such as PNG or JPEG, the fonts available
and the method for specifying new fonts will be the same as when we use a
screen graphics device, so the descriptions from the previous section apply.

PDF and PostScript fonts

The vector graphics devices for PDF and PostScript work the same across
all operating systems. This means that we may get a slightly different (text)
result in PDF output compared to what we see on a screen graphics device, but
the PDF result should look the same for anyone else on any other computer.

Both PDF and PostScript graphics devices make use of *Type1* fonts. A Type1
font consists of at least two separate files: one file contains the descriptions
of the individual characters in the font, and another file contains *font metric*
information—the ascent, descent, and width of individual characters. A new
Type1 font can be defined in R with the `Type1Font()` function by supplying
four or five font metric files. The first four files describe font metrics for
normal, bold, italic, and bold-italic font faces and the fifth, if given, describes
the font metrics for a symbol font face.

The symbol font face is used for drawing mathematical equations, separate
from whatever font is being used for other text (see Section 10.5). Graphics
devices provide a default symbol font, so this is not required when defining a
new Type1 font.

Once a Type1 font is defined, it must also be registered for use with PDF using the `pdfFonts()` function and/or PostScript with the `postscriptFonts()`
function. The following code provides an example of defining a new font, registering it for use with the PDF graphics device, and drawing text with that
new font (see Figure 10.6).

hello

Figure 10.6

Text drawn on a PDF graphics device using a custom font called "flubber".

```
> flubber <- Type1Font("flubber",
                       rep(file.path(getwd(), "Type1",
                                     "flubber.afm"), 4),
                       encoding="WinAnsi.enc")
> pdfFonts(flubber=flubber)
> pdf("flubber.pdf", width=4.5, height=.5)
> grid.rect(gp=gpar(col="gray"))
> grid.text("hello", gp=gpar(fontfamily="flubber"))
> dev.off()
> embedFonts("flubber.pdf", outfile="flubber-embedded.pdf",
             fontpaths=file.path(getwd(), "Type1"))
```

An extra complication with the PDF format is that the font does not have to be included in the PDF file. This means that, when we view a PDF file, if the computer we are viewing on does not have the required font, the PDF viewer can substitute a different font so that we can still see the text. However, the result is not usually very pleasant; the text may be readable, but it is usually ugly. To avoid this situation, we can embed the font as part of the PDF file. This is the purpose of the embedFonts() function call in the code above. If we embed the font using embedFonts() then the PDF file will look the same for someone else on another computer regardless of whether they have the font installed on their computer.

Type1 fonts are single-byte fonts, which means that the font can only include up to 256 different characters. It is not possible to include all characters and symbols from all languages in only 256 characters, so two different fonts may include two different sets of characters. This means that a Type1 font may also require an encoding file, which provides a list of the characters that are included in the font.

When we define a Type1 font, we can specify an encoding file for the font. This can be a complete path to an encoding file, or it can just be the name of one of the standard encoding files that comes included with R. For example, R provides the encoding file "ISOLatin1.enc", which includes, besides the basic ASCII english set of characters, a set of European accented characters, such as é.

Many modern fonts are available in *TrueType* or *OpenType* formats, but these can usually be converted to Type1 format for use with R.

The **extrafont** package, which was mentioned earlier in the context of using custom fonts on Windows, provides similar convenience when working with PDF and PostScript graphics devices. The fonts that are discovered by the font_import() function can be registered in bulk for use with PDF and PostScript simply by changing the argument to the loadfonts() function. For example, the following code registers all discovered system fonts for use with PDF output. The **extrafont** package actually only discovers TrueType fonts that have been installed, but it automatically takes care of converting the TrueType fonts to Type1 fonts.

```
> loadfonts(device="pdf")
```

SVG fonts

The SVG graphics device is a Cairo device (like the default screen device on Linux). This means that font selection is straightforward (as long as the required font is installed on our system). The SVG device also produces a result that is identical to other Cairo-based output. For example, on Linux, a plot on a screen device should look identical to the SVG result. The cost of this cross-format consistency is that text in SVG is rendered as paths. This means that the rendering of very small text on an svg() device will not look as good as, say, that on the pdf() device (because *hinting* information from the font, which tells renderers how to draw very small text, is lost when the font is converted to a path). The fact that text is just a path in svg() output also means that we cannot search within the text in the resulting SVG file.

This text behaviour is specific to the svg() graphics device; it is not a feature of SVG in general. For example, Chapter 13 describes some other ways to get SVG output from R for which the selection of fonts and the rendering of text is quite different.

Cairo fonts

The Cairo graphics system is behind several R graphics devices. We have previously mentioned the Cairo screen device that is the default on Linux (which also produces PNG and JPEG output on Linux) and the SVG graphics device.

In addition, there are Cairo versions of PDF and PostScript devices, which are available via the cairo_pdf() function and the cairo_ps() function. One advantage of using these graphics devices is that they support a wider range of

> | māori |
> | --- |

Figure 10.7
Text that includes a macron-accented character drawn on a Cairo PDF graphics device.

fonts (all fonts installed on the computer). These devices also automatically embed fonts. Furthermore, these Cairo graphics devices support UTF8 text, which means that they can draw any character or symbol from any language (as long as the font contains the character or symbol). For example, if we wish to draw an 'a' character with a macron accent, this is straightforward on a Cairo device. There is no need to register fonts or specify encodings; we can just name a font and we can just specify characters using Unicode escapes. The following code provides an example, with an "Ubuntu" font selected just by name (see Figure 10.7).

```
> cairo_pdf("cairo.pdf", width=4.5, height=.5)
> grid.rect(gp=gpar(col="gray"))
> grid.text("m\U0101ori", gp=gpar(fontfamily="Ubuntu"))
> dev.off()
```

On Windows, the default screen device, and hence the default raster device for PNG and JPEG output, is not based on Cairo. In this case, we can choose to use Cairo graphics for raster output via the `type` argument to the `png()` function. There is also a **Cairo** package to produce Cairo-based output on screen in Windows.

One reason for not using Cairo graphics for all output is that it will sometimes produce raster output even in vector formats (e.g., if the image involves semitransparency).

LATEX fonts

A particularly important set of fonts are the Computer Modern fonts that are the default in LATEX documents. If we are producing plots for inclusion in a LATEX document, we might like to use Computer Modern for the plot labels so that they match the main text.

The computer modern fonts are available in Type1 versions (e.g., the fonts-cmu package on Linux), so one approach is to install those and use the instructions above for using Type1 fonts with the `pdf()` device or with a Cairo-based device. The **fontcm** package (based on **extrafont**) simplifies this process significantly.

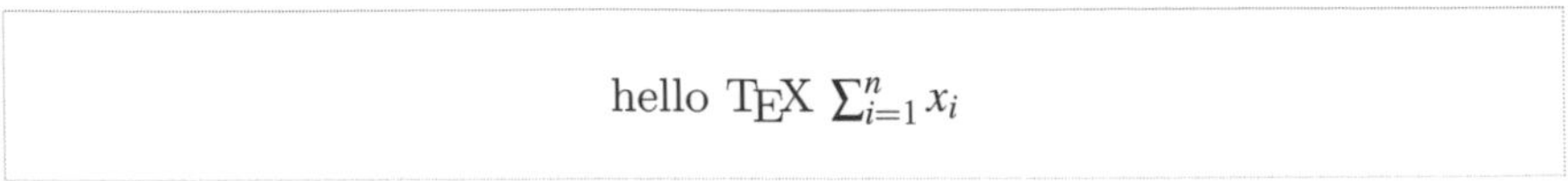

hello T_EX $\sum_{i=1}^{n} x_i$

Figure 10.8

R graphics output produced with the `tikz()` device.

Another option is to use the **tikzDevice** package, which provides a `tikz()` device. This uses T_EX (via pgf/TikZ) to produce graphics output, so natively makes use of L^AT_EX Computer Modern fonts. This graphics device also interprets T_EX commands within text, including mathematical equations. The following code provides a simple demonstration (see Figure 10.8).

```
> library(tikzDevice)
> tikz("params-tikz.tex", width=4.5, height=.5)
> grid.rect(gp=gpar(col="grey"))
> grid.text("hello \\TeX{} $\\sum_{i=1}^n x_i$")
> dev.off()
```

Symbol fonts

When we draw mathematical equations in a plot (Section 10.5), R makes use of a special *symbol font*, which contains the special characters required for mathematical equations, and ignores the current font selection for normal text.

Most graphics devices do not allow the user to choose a different symbol font, but it is possible to specify a custom font for Type1 fonts, for use with PDF or PostScript output. The complication with selecting a symbol font is that the font that we choose must obey the Adobe Symbol Encoding.

Setting the encoding for a font is not very easy, but in one particular case the hard work has been done for us. The **fontcm** package provides a symbol font, encoded for use with R, that is based on the Computer Modern math fonts, so that mathematical equations are closer in appearance to what L^AT_EX produces.

As mentioned in the section on L^AT_EX fonts, another way to produce R graphics with L^AT_EX fonts, including mathematical equations, is with the device from the **tikzDevice** package.

---

hello

Figure 10.9
R graphics text drawn with a Google Font (Special Elite) provided by the **showtext** package.

Package showtext

One final solution for selecting a font in R graphics is the **showtext** package. The advantages of this approach are that it provides access to the widest range of fonts, including Google Fonts, and it should work on all graphics devices. The downside is that text is drawn as paths rather than as a proper font (similar to what the `svg()` device does), which will produce a lower quality result in some situations (e.g., very small text). This approach also requires additional function calls to "activate" the **showtext** functionality. The following code shows a simple example: we choose a font with `font_add_google()`, mapping the name of the font we want to a font family name to use in R; we activate **showtext** rendering with `showtext_begin()`; we draw the text, using the font family name we defined in the first step; and we deactivate **showtext** with `showtext_end()`. The final result is shown in Figure 10.9.

```
> library(showtext)
> font_add_google("Special Elite", "elite")
> grid.rect(gp=gpar(col="grey"))
> showtext_begin()
> grid.text("hello", gp=gpar(fontfamily="elite"))
> showtext_end()
```

10.4.2 Font face

The font face is usually specified as an integer value between 1 and 4. Table 10.3 shows the mapping from numbers to font faces.

The **grid** graphics system also allows the font face to be specified by name (see Table 6.5).

R graphics is limited to only bold, italic, and bold-italic variations of a font, where other graphics systems provide a wider range of possibilities. For example, the CSS `font-stretch` property allows specification of "condensed" or "expanded" variants of a font.

Table 10.3

Possible integer font face specifications and their meanings. See Table 6.5 for font face *name* specifications. The range of valid font faces varies for different Hershey fonts, but the maximum valid value is usually 4 or less. When the font family is `"HersheySerif"`, there are a number of special font faces available.

Integer	Description
1	Roman or upright face
2	Bold face
3	Slanted or italic face
4	Bold and slanted face
5	Symbol

For the HersheySerif font family

5	Cyrillic font
6	Slanted Cyrillic font
7	Japanese characters

However, it is possible to work around the limitation in R by specifying a variation like condensed in the font family name. The following code provides a simple example, showing the normal Ubuntu font family, with font face variations, plus an Ubuntu Condensed font family (see Figure 10.10).

```
> cairo_pdf("cairo-faces.pdf", width=4.5, height=.5)
> grid.rect(gp=gpar(col="grey"))
> grid.text("hello", x=1/5,
            gp=gpar(fontfamily="Ubuntu"))
> grid.text("hello", x=2/5,
            gp=gpar(fontfamily="Ubuntu", fontface="bold"))
> grid.text("hello", x=3/5,
            gp=gpar(fontfamily="Ubuntu", fontface="italic"))
> grid.text("hello", x=4/5,
            gp=gpar(fontfamily="Ubuntu Condensed"))
> dev.off()
```

10.4.3   Multi-line text

It is possible to draw text that spans several lines by inserting a new line escape sequence, `"\n"`, within a piece of text, as in the following example.

```
"first line\n second line"
```

hello **hello** *hello* hello

Figure 10.10
R graphics text drawn with the Ubuntu font. The right-most text is drawn in a condensed font by specifying a font family that is a condensed font.

Alternatively, simply entering a character value across several lines will produce the same result, as shown below.

```
> "first line
    second line"
```

```
[1] "first line\n second line"
```

Vertical separation of the text for drawing can be controlled via a line height parameter, which acts as a multiplier (2 means double-spaced text).

10.4.4   Locales

R supports multibyte locales, such as UTF-8 locales and East Asian locales (Chinese, Japanese, and Korean), which means that it is possible to enter multibyte character values. There may be problems including such characters as part of graphical output on some devices. For example, Type1 fonts on PostScript and PDF devices only work with single-byte character encodings, so an appropriate encoding may need to be specified in order to produce special characters on those devices. The Cairo-based devices, `cairo_pdf()` and `cairo_ps()`, are a good way around this issue.

10.4.5   Escape sequences

Another issue is how to type characters that are not directly represented on our keyboard. One option is to use one of the escape sequences that R allows within character values. These include simple escape sequences for non-printing characters, such as "\n" for a newline and "\t" for a tab, octal sequences of the form "\nnn" (where n is an octal digit), e.g., "\351" for é (in an ISOLatin1 locale), hexadecimal sequences of the form "\xnn" (where n is a hex digit), e.g., "\xE9" for é (in an ISOLatin1 locale), and Unicode sequences of the form "\unnnn" (where n is a hex digit), e.g., "\u00E9" for é.

Figure 10.11
The text on the left has anti-aliasing turned off, the middle text has grayscale anti-aliasing, and the text on the right has subpixel anti-aliasing.

10.4.6   Anti-aliasing

The appearance of text, with its fine detail, is greatly improved by anti-aliasing—using, for example, grey pixels along the edge of character glyphs to make the edges appear smooth (see Figure 10.11). When we are producing vector output, such as PDF, we do not have to worry about this issue because it is handled by the viewer or printer that renders the PDF document. However, when we are producing raster output, such as PNG, or drawing to the screen, we may need to specify whether we want anti-aliasing applied, and if so, what sort.

The X Window device does not perform anti-aliasing of text, so there is nothing to do in this case. On Cairo-based devices, the `antialias` argument can be set to `"none"` for no anti-aliasing, `"gray"` to use grayscale pixels, and `"subpixel"` to use subpixel anti-aliasing. The default is selected based on the system that R is running on, and will typically be `"gray"`. Subpixel anti-aliasing means that individual red, green, and blue components of a pixel are adjusted to create a smooth edge (rather than using complete pixels). When it works well, subpixel anti-aliasing produces a smoother result than grayscale anti-aliasing, but subpixel anti-aliasing does not always work well on all types of display.

On Windows, for raster devices, the `antialias` argument can be set to `"none"`, `"gray"`, or `"cleartype"`, where the latter produces subpixel rendering. On MacOS X, the Quartz device always uses anti-aliasing unless `antialias="none"` is specified.

10.5 Mathematical formulae

This section does not concern a graphical parameter, but it does provide important information about how to specify character values for drawing text.

Any R graphics function that draws text should accept both a normal character value, e.g., `"some text"`, and an R expression, which is typically the result of a call to the `expression()` function. If an expression is specified as the text to draw, then it is interpreted as a mathematical formula and is formatted appropriately. This section provides some simple examples of what can be achieved. For a complete description of the available features, type `help(plotmath)` or `demo(plotmath)` in an R session.

When an R expression is provided as text to draw in graphical output, the expression is evaluated to produce a mathematical formula. This evaluation is very different from the normal evaluation of R expressions: certain names are interpreted as special mathematical symbols, e.g., `alpha` is interpreted as the Greek symbol α; certain mathematical operators are interpreted as literal symbols, e.g., a `+` is interpreted as a plus sign symbol; and certain functions are interpreted as mathematical operators, e.g., `sum(x, i==1, n)` is interpreted as $\sum_{i=1}^{n} x$. Figure 10.12 shows some examples of expressions and the output that they create.

Some of the operators within an expression affect the style of text and this provides a limited way to control the font face within a single piece of text. For example, the following code produces text that includes italic and bold words (see Figure 10.13).

```
> grid.text(expression("We can make text "*
                       italic("emphasized")*
                       " or "*
                       bold("strong")))
```

In some situations, for example, when calling graphics functions from within a loop, or when calling graphics functions from within another function, the expression representing the mathematical formula must be constructed using values within variables as well as literal symbols and constants. A variable name within an expression will be treated as a literal symbol (i.e., the variable name will be drawn, not the value within the variable). The solution in such cases is to use the `substitute()` function to produce an expression. The following code shows the use of `substitute()` to produce a label where the year is stored in a variable.

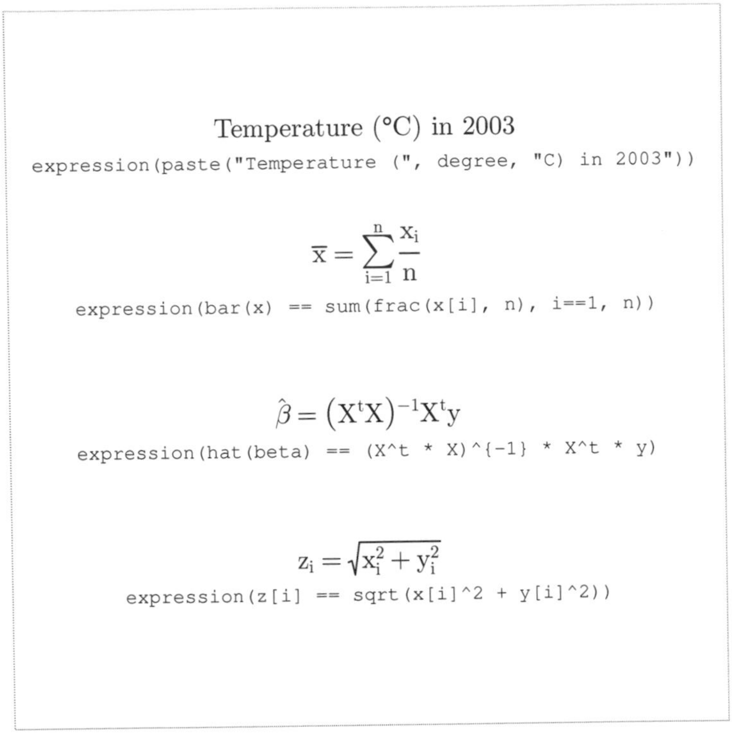

Temperature (°C) in 2003

```
expression(paste("Temperature (", degree, "C) in 2003"))
```

$$\bar{x} = \sum_{i=1}^{n} \frac{x_i}{n}$$

```
expression(bar(x) == sum(frac(x[i], n), i==1, n))
```

$$\hat{\beta} = \left(X^t X \right)^{-1} X^t y$$

```
expression(hat(beta) == (X^t * X)^{-1} * X^t * y)
```

$$z_i = \sqrt{x_i^2 + y_i^2}$$

```
expression(z[i] == sqrt(x[i]^2 + y[i]^2))
```

Figure 10.12
Mathematical formulae in plots. For each example, the output is shown in a serif font, and below that, in a typewriter font, is the R expression required to produce the output.

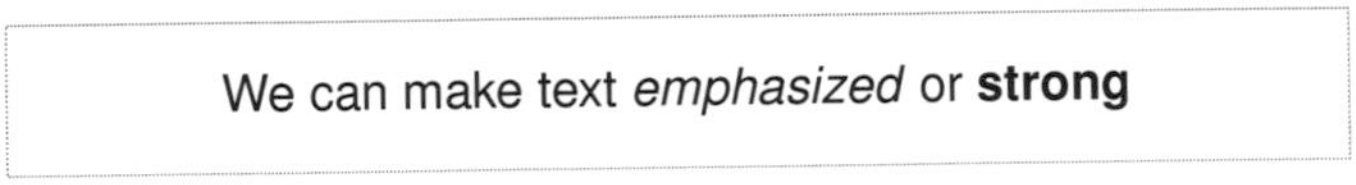

We can make text *emphasized* or **strong**

Figure 10.13
Specifying multiple font faces within the same piece of text, by using `italic()` and `bold()` within an expression (within a call to a text-drawing function).

```
> myfunction <- function(year) {
    text(0.5, 0.5, substitute(paste("Temperature (",
                                degree, "C) in ", year),
                        list(year=year)))
}
```

The mathematical annotation feature makes use of information about the dimensions of individual characters to perform the formatting of the formula. For some output formats, such information is not available, so mathematical formulae cannot be produced.

Chapter summary

There are standard ways to specify colors, fonts, line types, and text for virtually all graphics functions in R. There are functions for generating coherent sets of colors as well as individual colors. Specifying fonts for text can be tricky and format-dependent and platform-dependent. Text can be specified as an R expression, which makes it possible to draw special characters and to produce special formatting for mathematical formulae.

Part IV

INTEGRATING GRAPHICS SYSTEMS

11

Importing Graphics

Chapter preview

This chapter describes packages and functions that import images from external files and allow them to be included as part of R graphics output. There are separate packages for importing raster images and importing vector images.

Sections 3.4.1 and 6.2 described the set of graphical primitives that are available in the base graphics system and the **grid** graphics system. These graphical primitives make it possible to draw basic shapes, text, and bitmap images and they form the basis for drawing more complex images with R.

By combining basic shapes, it is possible to produce an infinite variety of pictures; however, there are still some images that cannot be produced with R, and R is not the best way to produce many kinds of images. For example, it is not possible to generate a photographic image with R and there are much better programs than R for producing artistic images such as logos.

Images like photographs and logos can be useful in plots or pictures, for example, to provide a background image for a plot, or to annotate a plot with the logo of a company or institution. In such cases, it may be necessary, or just more convenient, to create the image outside of R and *import* the image into R.

A number of packages provide tools for importing graphics into R and the choice of which one to use will depend on the format of the original image and what is to be done with the image once it has been imported. Image formats can be divided into *raster* formats and *vector* formats (see Section 9.2.1) and packages that import images into R typically address one of these options.

Figure 11.1
Three images of the Moon. On the left is a JPEG photograph of the North Pole of the
Moon that has been assembled from images taken by the Galileo spacecraft, courtesy
of NASA (`image #: PIA00130`). In the middle is a cartoon image of the Moon from
the Open Clip Art Library `http://openclipart.org/media/files/rg1024/10351`.
On the right is another vector image of the Moon from Pixabay `https://pixabay.`
`com/en/moon-planet-outer-space-26619/`.

11.1 The Moon and the tides

To provide a concrete example of importing images into R, this section looks
at producing a plot that shows the relationship between the timing of low tide
and the phase of the Moon. The main plot shows the hour during the day at
which low tide occurs as a function of the day of the month and the phases of
the Moon, and the plot is "dramatized" by adding an image of the Moon in
the background.

Three versions of this plot are considered: one using a raster image photograph
of the North Pole of the Moon, taken by NASA's Galileo spacecraft, one using
a relatively simple vector image from the Open Clip Art Library, and one
using a relatively complex vector image from Pixabay (see Figure 11.1). The
difference between the latter two is that the simpler vector image only uses
graphical primitives that R graphics supports, while the more complex vector
image uses a radial gradient fill that goes beyond the capabilities of standard
R graphics.

One version of the plot, using the raster moon image as background, is shown
in Figure 11.2. The complete code for this plot is available on the book's
web site. The focus of this chapter is on the two *conceptual* steps involved in
producing the plot: An external image has to be *read* by R; and the image
has to be *rendered* by R.

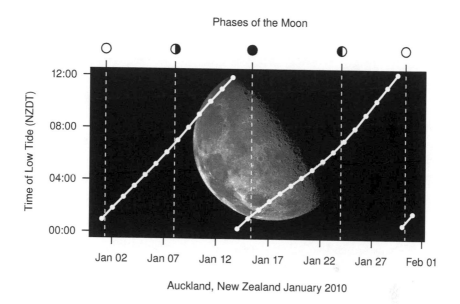

Figure 11.2

A plot with a raster background image. A raster photograph of the Moon provides a backdrop to a line plot of times of low tide for Auckland in January 2010 (data were obtained from Land Information New Zealand; `http://hydro.linz.govt.nz`).

Table 11.1
A selection of packages that can read external raster
images into R.

Package	Function	File Formats
png	readPNG()	PNG
jpeg	readJPEG()	JPEG
tiff	readTIFF()	TIFF
magick	image_read()	Multiple formats

11.2   Importing raster graphics

Examples of raster formats include JPEG and PNG. A standard source of JPEG files are photographs from digital cameras, while PNG images are commonly encountered on the web.

The first step is to find a function that can read the file format of the external image. A large number of packages provide functions for reading image formats and Table 11.1 describes some of these. The important differences between these functions are the range of file formats that they can handle and how much they depend on other software (i.e., how much other software must also be installed).

The **magick** package provides an R interface to the ImageMagick image manipulation system. This means that it can read images in many different formats. The downside is that we need ImageMagick installed on our system for this package to work.

The `dev.capture()` function can also be used to capture the current R graphics device as a raster image (for raster-based devices, such as screen devices).

Having read an image into R, the next step is to draw the image as part of an R plot. In base graphics, this means using the `rasterImage()` function (see Section 3.4.1), and in **grid** this means using `grid.raster()` (see Section 6.2) because those functions allow the image to be drawn relative to the plot regions and coordinate systems of an R plot (for example, Figure 11.2).

For the packages listed in Table 11.1, the result of reading an external file is an R object that can be passed immediately to `rasterImage()` or `grid.raster()`.

For our example of a line plot with a moon image in the background, the original image is a JPEG file, so the following code can be used to read the image into R.

```
> library(jpeg)
> moon <- readJPEG(system.file("extra", "GPN-2000-000473.jpg",
                        package="RGraphics"))
```

The result, `moon`, is a `"matrix"` object, so drawing the image with **grid** is as simple as the following code.

```
> grid.raster(moon)
```

However, that just draws the image as large as it can go on the current page. We can position and size the image with arguments to `grid.raster()` or by calling `grid.raster()` within an appropriate **grid** viewport.

The following code draws the moon image in the top-left corner of the page, specifying only its height so that it retains its aspect ratio, then in the top-right corner at a fixed size that distorts the image, then in a series of viewports across the bottom of the page at different angles (see Figure 11.3).

```
> grid.raster(moon, x=0, y=1, height=.5, just=c("left", "top"))
> grid.raster(moon, x=1, y=.75,
              width=.5, height=.25, just="right")
> for (i in seq(10, 90, 10)) {
    pushViewport(viewport(x=i/100, y=.25, width=.2, height=.2,
                        angle=i - 10))
    grid.raster(moon)
    popViewport()
  }
```

The following code gives an indication of how we might position the image within a base plot. This is a segment of the code that was used to produce Figure 11.2; the full code is available on the book's web site.

```
> plot(lowTideDate, lowTideHour, xlab="Date", ylab="Time of Day")
> width <- grconvertX(1.5, "in", "user")
> aspect <- nrow(moon)/ncol(moon)
> height <- grconvertY(1.5*aspect, "in", "user")
> rasterImage(moon, usr[1], usr[3] + (usr[4] - height),
              width, usr[4])
```

The important point about this code is that we are able to locate the raster image relative to the plot regions and coordinate systems of the R plot and

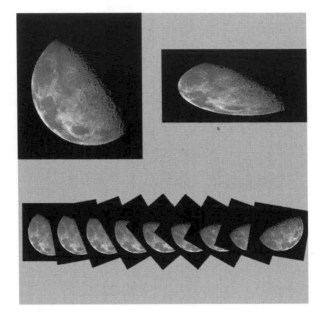

Figure 11.3
Drawing an imported raster image with **grid**. We can locate and size the raster image using arguments to `grid.raster()` (top-left and top-right) or by drawing within an appropriate viewport (bottom row).

we can do this programmatically. This is much better than trying to combine the R plot and the raster image using other software that has no access to the R plot regions and coordinate systems, or cannot be controlled programmatically.

The downside is that it may be necessary to perform some calculations to retain the aspect ratio of the image within the plot. In this case, we used the functions grconvertX() and grconvertY() to help out. This is much less of a problem when combining a raster image with a **grid**-based plot using grid.raster() because **grid** provides access to more useful coordinate systems (see Section 6.3).

If a package does not create an R object that we can use with rasterImage() or grid.raster(), we may have to convert the image to a matrix or array.

11.3 Importing vector graphics

Examples of vector image formats include PDF, PostScript, and SVG. In this section we will look at the **grImport** package, which can import PostScript images, and the **grImport2** package, which can import SVG images.

11.3.1 The grImport package

There are at least three steps required to draw an external PostScript image in an R plot. In the simplest case, where the original image is already a PostScript image, the first step is to "trace" the PostScript image using the PostScriptTrace() function.

The PostScriptTrace() function converts a PostScript image into an XML format (using Ghostscript). This step only needs to be performed once for each image. For example, the following code traces a PostScript format cartoon moon image (the middle image from Figure 11.1).

```
> library(grImport)
> PostScriptTrace(system.file("extra", "comic_moon.ps",
                              package="RGraphics"),
                 "comic_moon.xml")
```

The result of tracing is an XML file, and the second step involves reading that XML file into R using the readPicture() function.

```
> vectorMoon <- grImport::readPicture("comic_moon.xml")
```

The third step is to render the object created by `readPicture()`, in this case
`vectorMoon`, either using the `picture()` function, which will draw the image
in the current base graphics plot region, or using the `grid.picture()` function, which will draw the image in the current **grid** viewport. The following
code shows `grid.picture()` being used to draw the `vectorMoon` in several
different ways: the default is to fill the current viewport (with a small margin
around all sides); we can control the location and size of the image, by default
retaining the default aspect ratio; or we can deliberately distort the image
(see Figure 11.4).

```
> grImport::grid.picture(vectorMoon)
> grImport::grid.picture(vectorMoon,
                         x=0, y=1, just=c("left", "top"),
                         width=.2, height=.2)
> grImport::grid.picture(vectorMoon,
                         x=1, y=1, just=c("right", "top"),
                         width=.3, height=.1, distort=TRUE)
```

Figure 11.5 shows a more complex embedding of the vector moon image within
a **grid** viewport (the data region of a plot). The following code draws a simpler
version of that plot (the full code for Figure 11.5 is available from the book's
web site).

```
> library(lattice)

> xyplot(lowTideHour ~ lowTideDate, pch=16, col="black",
         xlab="Date", ylab="Time of Day",
         panel=function(x, y, ...) {
             grid.picture(vectorMoon)
             panel.xyplot(x, y, ...)
         })
```

Complications with importing vector images

An extra step is required (at the start) if the original image is not in PostScript
format. In that case, we need to use another software tool to convert the image
to PostScript. The **magick** package has an `image_convert()` function that
can perform this task. The Inkscape image drawing software* produces good
results for converting from an SVG image.

---

*`http://inkscape.org/`.

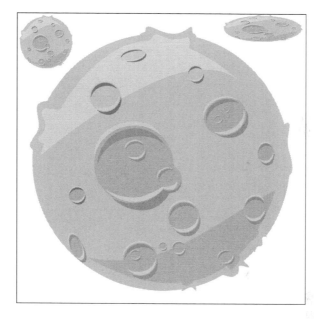

Figure 11.4
Drawing an imported vector image with **grid**. The image fills the current viewport by default, or we can locate and size the image using arguments to `grid.picture()` (top-left and top-right).

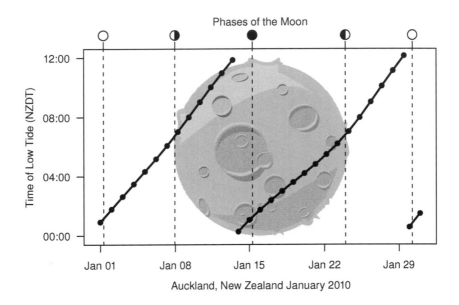

Figure 11.5
A plot with a vector background image. A vector cartoon of the Moon provides a backdrop to a line plot of times of low tide for Auckland in January 2010 (data were obtained from Land Information New Zealand; `http://hydro.linz.govt.nz`).

One danger is that, for complex vector images, some tools may convert the image, or parts of it, to a raster format. A number of other issues may also arise, mainly due to the fact that the content of a vector image can vary much more than the content of a raster image.

A raster image can be thought of as simply a two-dimensional array of pixels. There are many different ways that an array of pixels can be stored in a file, but the image structure is fundamentally always the same and very simple. This means that there are very few variations on how to read a raster image into R or how to draw a raster image as part of an R plot. The functions to read and draw raster images have relatively few arguments.

By contrast, a vector image is made up of a number of shapes or *paths*. There may be very few paths, or very many paths. The paths may overlap each other or even intersect with themselves. There may be text (letters are essentially quite detailed and complex paths) and, in more complex cases, one path may be used just to define a clipping region and not be drawn at all.

Sometimes, these complications mean that R will not be able to import an image or it may not render the original image properly. In any case, reading in a vector image and rendering the image may require more than a single step. In particular, it may be necessary to work with individual paths within a vector image and the **grImport** package provides several tools for doing so.

Manipulating vector images

One convenient feature is the ability to subset the object that is created by the **readPicture()** function. For example, the following code just draws the first four paths in the image (see Figure 11.6).

```
> grImport::grid.picture(vectorMoon[1:4])
```

There is also a **picturePaths()** function that allows each path to be inspected in isolation. The following code shows the first six paths within the cartoon Moon image (see Figure 11.7).

```
> grImport::picturePaths(vectorMoon[1:6], fill="white",
              freeScales=TRUE, nr=2, nc=3)
```

It is also useful to note that the imported image is essentially just a series of polygon outlines. The following code draws a "wireframe" version of the Moon image by ignoring the colors from the original image and just drawing the outline of each path (see Figure 11.6).

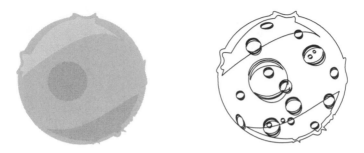

Figure 11.6

On the left, a "subset" of the cartoon Moon image (Figure 11.1), consisting of only the first four paths. On the right, the paths from the cartoon Moon image drawn as simple outlines, ignoring the fill colors from the original image.

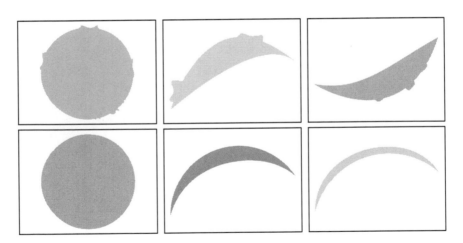

Figure 11.7

The first six paths (shapes) in the cartoon Moon image from Figure 11.1.

```
> grImport::grid.picture(vectorMoon, use.gc=FALSE)
```

These facilities can be used, for example, to exclude certain parts of an image, or to render paths in a different order, which can sometimes be useful to reproduce the original image faithfully with R graphics.

Imported images as data symbols

The **grImport** package provides a convenience function, `grid.symbols()`, that can be used to draw an imported image at multiple locations with a single function call. For example, the following code draws a vector moon image at each data point in a scatterplot.

```
> xyplot(lowTideHour ~ lowTideDate, pch=16, col="black",
        xlab="Date", ylab="Time of Day", subset=1:10,
        panel=function(x, y, ...) {
            grid.symbols(vectorMoon, x, y, units="native",
                         size=unit(10, "mm"))
        })
```

Drawing with this function will become very slow with a large number of data points.

11.3.2 The grImport2 package

The **grImport2** package is similar to the **grImport** package, except that it imports SVG images rather than PostScript images. One obvious use for this package is when the original image is in an SVG format. For example, the PostScript cartoon moon image from the previous section was originally an SVG image. With **grImport2**, rather than converting it to PostScript for importing with **grImport**, we can import the original SVG image directly.

As with **grImport**, there are at least three steps involved in importing an image with **grImport2**: pre-processing the image, reading the image into R, and rendering the image in R.

The **grImport2** package is able to directly read SVG files, but the SVG file must have been generated by the Cairo graphics library. This means that we have to use a program that generates Cairo SVG to pre-process the SVG file. The `rsvg_svg()` function from the **rsvg** package is one way to perform this step. For the case of the cartoon moon, the code is shown below. This code starts with the original SVG file, `"comic_moon.svg"`, which was not created

Figure 11.8
A **grImport2** rendering of the cartoon moon.

by Cairo graphics, and generates a new SVG file, `"comic_moon_cairo.svg"`, which describes the same image, but can be read by **grImport2**.

```
> library(rsvg)
```

```
> rsvg_svg(system.file("extra", "comic_moon.svg",
           package="RGraphics"),
           "comic_moon_cairo.svg")
```

The resulting Cairo SVG file can then be read into R with the `readPicture()` function from **grImport2**.

```
> library(grImport2)
> moonSVG <- grImport2::readPicture("comic_moon_cairo.svg")
```

Finally, the resulting R object, `moonSVG`, can be drawn with the function `grid.picture()` from **grImport2** (there is no support for drawing the R object with base graphics). The result of the following code is shown in Figure 11.8.

```
> grImport2::grid.picture(moonSVG)
```

The `grid.picture()` function from **grImport2** works the same way as the same function from **grImport**; the imported image is drawn in the current **grid** viewport, filling the viewport (with a small margin on all sides) and

Figure 11.9
A test image for importing. The image consists of a gray rectangle and a line of text, with the text obscured behind a simple raster image.

retaining the original image aspect ratio by default. Alternatively, we can specify a location and size for the imported image within the viewport if we wish.

The main advantage of using **grImport2** is that it allows us to import more complex and sophisticated images than **grImport**. For example, Figure 11.9 shows a simple test image consisting of a gray rectangle and a line of text, with the text obscured by a simple raster image.

The following output shows the grobs in this image.

```
> grid.ls(full.names=TRUE)
```

```
GRID.rect.1570
GRID.text.1571
GRID.rastergrob.1572
```

If we import this image using **grImport**, that package is not capable of importing the raster component of the image so if we render the image, the raster is missing and the line of text is visible (see Figure 11.10).

```
> PostScriptTrace("importtest.ps", "importtest.xml")
> test <- grImport::readPicture("importtest.xml")
> grImport::grid.picture(test)
```

The **grImport2** package does import raster elements, amongst other things, so it renders the test image correctly (see Figure 11.11).

Figure 11.10
The test image from Figure 11.9 as produced by **grImport**. The raster element of
the test image does not get imported, so we are able to see the text.

```
> rsvg_svg("importtest.svg", "importtest-cairo.svg")
> test <- grImport2::readPicture("importtest-cairo.svg")
> grImport2::grid.picture(test)
```

An interesting feature of **grImport2** is that it will even import features of
an image that R graphics does not support. The next example demonstrates
this idea. We have another vector moon image (see the right-hand image in
Figure 11.1), but this time the image consists not only of several filled paths,
but also several radial gradients.

To import this image, we follow the same steps as before, pre-processing the
image with `rsvg_svg()`, then reading the resulting Cairo-based SVG into R
with `grImport2::readPicture`. This reads all features of the image into R,
including the radial gradients.

```
> rsvg_svg(system.file("extra", "moon-26619.svg",
                       package="RGraphics"),
          "full-moon.svg")
> moon <- grImport2::readPicture("full-moon.svg")
```

If we draw the imported image on a normal R graphics device, we only see
the features of the image that R graphics supports; in this case, just a set of
filled paths (see Figure 11.12).

```
> grImport2::grid.picture(moon)
```

Figure 11.11

The test image from Figure 11.9 as produced by **grImport2**. The raster element of the test image does get imported (and we are not able to see the text).

Figure 11.12

The complex vector moon image drawn using a normal R graphics device. The filled paths component of the image has been rendered, but the radial gradient component has not.

Figure 11.13

The complex vector moon image drawn using **gridSVG**. The filled paths component of the image has been rendered *and* the radial gradient component of the image has been rendered.

Although R graphics does not support some SVG special effects, all features of an SVG image are imported by **grImport2**. This means that, if we use the **gridSVG** package (see Chapter 13) to produce SVG output, we can reproduce the imported SVG features in the exported SVG.

The following code demonstrates this idea. First, we load the **gridSVG** package. Then we call `grid.picture()` with the same imported R object as before, `moon`, but this time we specify `ext="gridSVG"`. This tells `grid.picture()` to make use of the features of the **gridSVG** package when rendering the imported image. Finally, we call `grid.export()` from the **gridSVG** package to generate an SVG file that contains all of the original features of the original SVG image (see Figure 11.13).

```
> library(gridSVG)

> grImport2::grid.picture(moon, ext="gridSVG")
> grid.export("moon3gridsvg.svg")
```

Figure 11.14 shows a more complex embedding of the vector moon image within a **grid** viewport (the data region of a plot). The following code draws a simpler version of that plot (the full code for Figure 11.14 is available from the book web site).

```
> xyplot(lowTideHour ~ lowTideDate, pch=16, col="white",
         xlab="Date", ylab="Time of Day",
         panel=function(x, y, ...) {
             grid.rect(gp=gpar(fill="black"))
             grImport2::grid.picture(moon, ext="gridSVG")
             panel.xyplot(x, y, ...)
         })
```

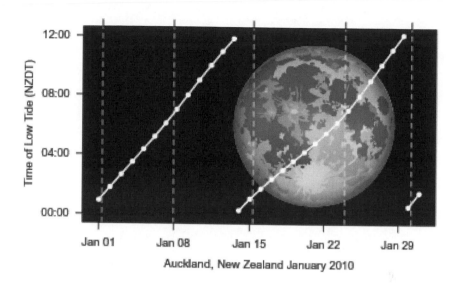

Figure 11.14
A plot with a complex vector background image that contains radial gradients. The vector Moon provides a backdrop to a line plot of times of low tide for Auckland in January 2010 (data were obtained from Land Information New Zealand; `http://hydro.linz.govt.nz`).

```
> grid.export("moon-plot.svg")
```

The **grImport2** package also has a `grid.symbols()` function for using an imported image as a data symbol, similar to **grImport**.

Although **grImport2** provides additional features that go beyond what the **grImport** package can do, the **grImport** package is still useful for several reasons. If the original image is PostScript, then we do not need to convert it to another format and risk losing or messing up elements of the image. The **grImport** package will also import text as text rather than always converting text to paths.

Chapter summary

A number of packages provide functions for reading raster images into R. The images can be drawn using base or **grid** graphical primitives. The **grImport** package provides functions for reading PostScript vector images into R and drawing them. The **grImport2** package provides functions for reading SVG vector images into R and drawing them.

12

Combining Graphics Systems

Chapter preview

This chapter describes the **gridBase** and **gridGraphics** packages, both of which make it possible to combine the output from the base graphics system with the output from the **grid** graphics system.

The **grid** graphics system and the base graphics system work completely independently of each other. This means that, while it is possible to produce output from both systems on the same page, there should normally be no expectation that the output from the two systems will correspond in any sensible way.

The **grid** graphics system offers more power and flexibility than the base graphics system, and the **lattice** and **ggplot2** packages provide some facilities not available in the base graphics system. However, it is often necessary to use the base system because many plotting functions in extension packages for R are built on the base system. Clearly, a combination of the wide range of base plots and the power and flexibility of **grid**, **lattice**, and **ggplot2** would be desirable. That task is the focus of this chapter.

This chapter describes two packages, **gridBase** and **gridGraphics**, both of which provide functions that can be used, in some situations, and with a little care, to overcome this inherent incompatibility and combine the output from the two systems in a coherent manner.

A discussion of the relative strengths and weaknesses of these packages will be delayed until the end of the chapter.

12.1   The gridBase package

12.1.1   Annotating base graphics using grid

The **gridBase** package has one function, `baseViewports()`, that supports
adding **grid** output to a base graphics plot. This function creates a set of
grid viewports (see Section 6.5) that correspond to the current base plot
regions (see Section 3.1.1). By pushing these viewports, it is possible to do
simple annotations to a base plot, such as adding lines and text using **grid**'s
units to locate them relative to a wide variety of coordinate systems, or to
attempt more complex annotations involving pushing further **grid** viewports.

The `baseViewports()` function returns a list of three grid viewports. The
first corresponds to the base graphics inner region. This viewport is relative
to the entire device and it only makes sense to push this viewport from the
"top level" (i.e., only when no other **grid** viewports have been pushed). The
second viewport corresponds to the base graphics figure region and is relative
to the inner region, and it only makes sense to push it after the inner viewport
has been pushed. The third viewport corresponds to the base graphics plot
region and is relative to the figure region, and it only makes sense to push it
after the other two viewports have been pushed in the correct order.

A simple application of this facility involves adding text to the margins of
a base graphics plot at an arbitrary orientation. The base graphics function
`mtext()` allows text to be located in terms of a number of lines away from
the plot region, but only at rotations of 0 or 90 degrees. The base graphics
`text()` function allows arbitrary rotations, but only locates text relative to
the user coordinate system in effect in the plot region (which is inconvenient
for locating text in the margins of the plot). By contrast, the **grid** function
`grid.text()` allows arbitrary rotations and can be used in any **grid** viewport.
In the following, a base graphics plot is created with the x-axis tick labels left
off (see Figure 12.1).*

```
> library(zoo)
> m <- factor(months(as.yearmon(time(sunspots)))),
              levels=month.name)
> plot(m, sunspots, axes=FALSE)
> axis(2)
> axis(1, at=1:12, labels=FALSE)
```

---

*This example uses data on the monthly mean relative sunspot numbers from 1749 to
1983, available as the **sunspots** data set in the **datasets** package.

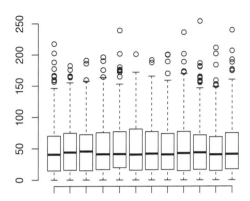

Figure 12.1
A base plot with no x-axis labels. Tick labels will be added at an angle of 60 degrees
using **gridBase** (see Figure 12.2).

In the next code, `baseViewports()` is used to create grid viewports that
correspond to the base graphics plot and those viewports are pushed.

```
> library(gridBase)
> vps <- baseViewports()
> pushViewport(vps$inner, vps$figure, vps$plot)
```

Finally, rotated labels are drawn using `grid.text()` (and the viewports are
popped to clean up). The final output is shown in Figure 12.2. The labels
can be located horizontally relative to the plot x-axis scale, so that they line
up with the plot tick marks, because the `vps$plot` viewport that we pushed
has the same axis scales as the base plot region (and the viewport is located
on the page in exactly the same position as the base plot region).

```
> grid.text(month.name,
            x=unit(1:12, "native"), y=unit(-1, "lines"),
            just="right", rot=60)
> popViewport(3)
```

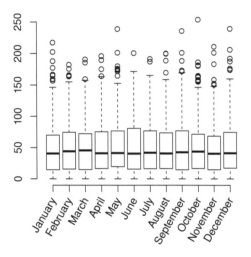

Figure 12.2

Annotating a base plot with **grid**. This is the plot from Figure 12.1 with the x-axis labels drawn using `grid.text()` to make use of both a convenient coordinate system (lines of text away from the x-axis) and the ability to rotate text to any angle.

12.1.2 Base graphics in grid viewports

The **gridBase** package provides several functions for adding base graphics output to **grid** output. There are three functions that allow base graphics plotting regions to be aligned with the current **grid** viewport: gridOMI(), gridFIG(), and gridPLT(). These make it possible to draw one or more base graphics plots within a **grid** viewport. The fourth function, gridPAR(), provides a set of graphical parameter settings so that base graphics par() settings can be made to correspond to some of the current **grid** graphical parameter settings.

The three functions return the appropriate par() values for setting the base graphics inner, figure, and plot regions, respectively.

The main usefulness of these functions is to allow the user to create a complex layout using **grid** and then draw a base graphics plot within relevant elements of that layout. The following example uses this idea to create a **lattice** plot where the panels contain dendrograms drawn using base graphics functions.

The first step just involves preparing some data to plot. A dendrogram object is created and cut into four subtrees.*

```
> hc <- hclust(dist(USArrests), "ave")
> dend1 <- as.dendrogram(hc)
> dend2 <- cut(dend1, h=70)
```

Next, some dummy-variables are created that correspond to the four subtrees.

```
> x <- 1:4
> y <- 1:4
> height <- factor(round(sapply(dend2$lower,
                                 attr, "height")))
```

Now a **lattice** panel function is defined to draw the dendrograms. The first thing this panel function does is push a viewport that is smaller than the viewport **lattice** creates for the panel. The purpose of this is to ensure that there is enough room for the labels on the dendrogram. The space variable contains a measure of the length of the longest label. The panel function then calls gridPLT() and uses the result in a call to par() to make the base graphics plot region correspond to the viewport that has just been pushed. It also sets new=TRUE so that the next call to plot(), when the next panel is

---

*This example uses data on violent crimes in the United States, available as the USArrests data set in the **datasets** package.

drawn, does not start a new page. Finally, the base `plot()` function is used to draw the dendrogram (and then the viewports that the panel function pushed are popped).

```
> space <- 1.2 * max(stringWidth(rownames(USArrests)))
> dendpanel <- function(x, y, subscripts, ...) {
    pushViewport(viewport(gp=gpar(fontsize=8)),
                 viewport(y=unit(0.95, "npc"), width=0.9,
                          height=unit(0.95, "npc") - space,
                          just="top"))
    par(plt=gridPLT(), new=TRUE, ps=8)
    plot(dend2$lower[[subscripts]], axes=FALSE)
    popViewport(2)
  }
```

Now the main plot can be drawn, using **lattice** to set up the arrangement of panels and strips (**grid** viewports) and the panel function defined above to draw a base graphics dendrogram in each panel.

```
> library(lattice)
```

The final plot is produced by a call to the `xyplot()` function (see Figure 12.3).

```
> plot.new()
> print(xyplot(y ~ x | height, subscripts=TRUE,
               xlab="", ylab="",
               strip=strip.custom(style=4),
               scales=list(draw=FALSE),
               panel=dendpanel),
        newpage=FALSE)
```

One difficulty with using **gridBase** to combine base and **grid** graphics output on the same page is that the two systems can end up fighting over who gets to start the drawing on a new page. In the code above, there is a call to `plot.new()` before the call to `xyplot()`. It is generally a good idea to start the new page with a call to `plot.new()` like this, rather than with `grid.newpage()`, or a high-level **lattice** or **ggplot2** function, because the **grid**-based functions tend to be more accepting of the fact that there may already be other drawing on the page.

This also explains the explicit call to `print()` around the `xyplot()` call, so that the `newpage` argument can be used to prevent `xyplot()` from starting its own new page. The general rule is: start a base plot and then add **grid** output to it rather than the other way around.

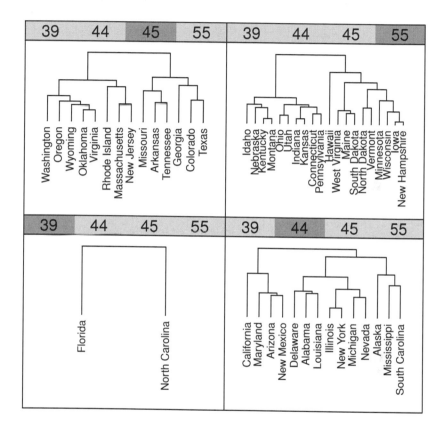

Figure 12.3

Embedding a base plot within **lattice** output. The arrangement of the panels and the drawing of axes and strips is all done by **lattice** using **grid**, but the contents of each panel is a dendrogram plot produced by the base graphics system.

12.1.3 Problems and limitations of gridBase

The functions provided by the **gridBase** package allow the user to mix output from two quite different graphics systems, but there are limits to how much the systems can be combined.

For example, it is not possible to embed base graphics output within a **grid** viewport that is rotated. There are also certain base graphics functions that modify settings like `omi` and `fig` themselves (e.g., `coplot()`) and output from these functions will not embed properly within **grid** viewports. Finally, the calculations used to match **grid** graphics settings with base graphics settings (and vice versa) are only valid for the current graphics device size. If these functions are used to draw into a window, then the window is resized, the base graphics and **grid** settings will almost certainly no longer match and the graph may become non-sensical. This also applies to copying output between devices of different sizes.

The `recordGraphics()` function provides one way to avoid this problem, though proper use of the function requires expert knowledge. A very naive use is shown in the following code.

```
> plot.new()
> recordGraphics({ print(xyplot(y ~ x | height,
                                 subscripts=TRUE,
                                 xlab="", ylab="",
                                 strip=strip.custom(style=4),
                                 scales=list(draw=FALSE),
                                 panel=dendpanel),
                           newpage=FALSE)
                },
                list(),
                globalenv())
```

Some other solutions to this problem are discussed in Section 7.13.

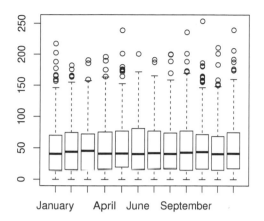

Figure 12.4
A base graphics boxplot, similar to Figure 12.1, but with x-axis labels included. Not all x-axis labels are visible because labels that overlap are not rendered.

12.2 The gridGraphics package

The **gridGraphics** package takes a very different approach to combining base and **grid** output. This package has a single main function, `grid.echo()`, and the purpose of this function is to convert a base plot into equivalent **grid** output.

For example, consider the base graphics boxplot from the previous section (Figure 12.1). We could draw the original barplot complete with labels on the x-axis using the following code (see Figure 12.4).

```
> plot(m, sunspots)
```

We can convert this to a **grid** plot by calling the `grid.echo()` function.

```
> library(gridGraphics)
```

```
> grid.echo()
```

The result is exactly the same as Figure 12.4, except that the plot has now been drawn using **grid**. We can see that by listing the **grid** grobs.

```
> grid.ls()
```

```
graphics-plot-1-polygon-1
graphics-plot-1-segments-1
graphics-plot-1-points-1
graphics-plot-1-segments-2
graphics-plot-1-segments-3
graphics-plot-1-polygon-2
graphics-plot-1-segments-4
graphics-plot-1-points-2
graphics-plot-1-polygon-3
graphics-plot-1-segments-5
...
```

12.2.1 Editing base graphics using grid

Now that we have **grid** grobs to work with, we can use the **grid** functions for editing grobs to change the position and orientation of the barplot labels. For example, the following code shifts the labels up closer to the x-axis, makes them right-justified, and rotates them 60 degrees (see Figure 12.5).

```
> grid.edit("graphics-plot-1-bottom-axis-labels-1",
            y=unit(-1, "lines"), hjust=1, vjust=0.5, rot=60)
```

12.2.2 Base graphics in grid viewports

By default the `grid.echo()` function just converts whatever base graphics output is on the current page. However, we can also provide a function as the first argument to `grid.echo()` and it will convert whatever base graphics output is produced by that function. This can be combined with the `newpage` argument to `grid.echo()` to draw base graphics within **grid** viewports; the base graphics is drawn on an off-screen device the size of the current **grid** viewport and then echoed (as **grid** output) into the current viewport.

As an example, we will revisit the example that we used for **gridBase**, to embed base graphics within a **lattice** plot (Figure 12.3). In the code below, the `dendpanel()` function is quite similar to the `dendpanel()` function that we used with **gridBase**, but instead of calls to `par()` and `gridPLT()` to set up

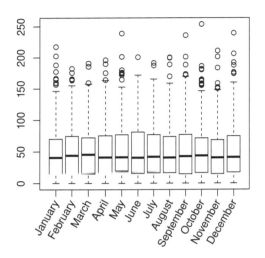

Figure 12.5
The base barplot from Figure 12.4, converted to a **grid** version and then edited
using **grid.edit()** to modify the x-axis labels.

base graphics plot regions plus a direct call to plot() to draw the dendrogram,
we just call grid.echo() and give it a function that calls plot() to draw the
dendrogram. We also specify newpage=FALSE so that grid.echo() just draws
in the **lattice** panel viewport. The prefix argument is used to provide names
for the **grid** grobs that are produced.

```
> dendpanel <- function(x, y, subscripts, ...) {
    pushViewport(viewport(gp=gpar(fontsize=8)),
                 viewport(y=unit(0.95, "npc"),
                          height=unit(0.95, "npc"),
                          just="top"))
    grid.echo(function() {
                par(mar=c(5.1, 0, 1, 0))
                plot(dend2$lower[[subscripts]], axes=FALSE)
              },
              newpage=FALSE,
              prefix=paste0("dend-", panel.number()))
    popViewport(2)
  }
```

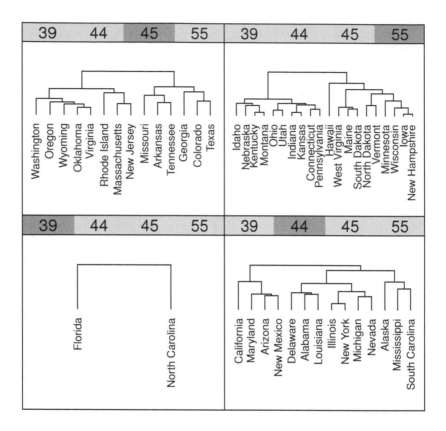

Figure 12.6
A **lattice** multi-panel plot with panels drawn by base graphics functions using
`grid.echo()`.

The call to `xyplot()` is exactly as it was in the **gridBase** case; we just provide
`dendpanel()` as the panel function in the call to `xyplot()`, although we no
longer have to worry about calling `plot.new()`. The final result is shown in
Figure 12.6.

```
> xyplot(y ~ x | height, subscripts=TRUE,
          xlab="", ylab="",
          strip=strip.custom(style=4),
          scales=list(draw=FALSE),
          panel=dendpanel)
```

12.2.3 Problems and limitations of gridGraphics

The **gridGraphics** package suffers from a few limitations, some of which are similar to the **gridBase** package. For example, the output of `grid.echo()` will not necessarily survive a device resize (e.g., resizing an on-screen graphics window or copying from one graphics device to another). The **gridGraphics** package also has limitations of its own. For example, it cannot exactly reproduce the labelling on a `contourplot()` and in some cases may censor axis labels in a slightly different manner than the base `axis()` function.

On the other hand, **gridGraphics** will perform better than **gridBase** in some respects. For example, it is possible with **gridGraphics** to combine base plots that usually demand the entire page for themselves, like the `coplot()` function. The following code demonstrates this idea by combining a `coplot()` with a **ggplot2** histogram on the same page (see Figure 12.7). We first define a function, `cpfun`, that contains a call to `coplot()`. Next, we push a viewport occupying the bottom 70% of the page and call `grid.echo()`, giving it the function `cpfun` (with `newpage=FALSE`) so that it draws a **grid** version of the conditioning plot in the bottom 70% of the page. The remainder of the code draws a **ggplot2** plot in a viewport that occupies the top third of the page.

```
> cpfun <- function() {
      coplot(lat ~ long | depth, quakes, pch=16, cex=.5,
          given.values=rbind(c(0, 400), c(300, 700)))
  }
> pushViewport(viewport(y=0, height=.7, just="bottom"))
> grid.echo(cpfun, newpage=FALSE, prefix="cp")
> upViewport()
> library(ggplot2)
> pushViewport(viewport(y=1, height=.33, just="top"))
> gg <- ggplot(quakes) + geom_histogram(aes(x=depth)) +
        theme(axis.title.x = element_blank())
> print(gg, newpage=FALSE)
> upViewport()
```

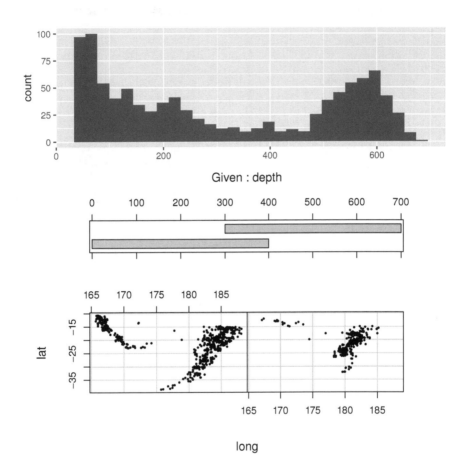

Figure 12.7
A `coplot()` conditioning plot (which usually demands the whole page for itself)
combined with a **ggplot2** histogram on the same page using `grid.echo()`.

Chapter summary

The **gridBase** package provides functions for aligning **grid** viewports with base graphics plot regions. This makes it possible to draw **grid**-based output within a base plot and base graphics output within **grid** viewports, including **lattice** and **ggplot2** plots.

The **gridGraphics** package provides the `grid.echo()` function, which converts base graphics output into **grid** graphics output. This also allows base plots to be drawn within **grid** viewports, plus it allows **grid**-style manipulation of base plots, with functions like `grid.edit()`.

13

Advanced Graphics

Chapter preview

The emphasis in this chapter is on accessing graphical effects and features that the core R graphics system does not support. The focus is on the **gridSVG** package for exporting R graphics to SVG with access to advanced features of the SVG format.

The strength of the core R graphics engine lies in the production of complex static plots with flexible control of fine details. However, there are limits to what the R graphics system will allow. For example, R graphics does not provide support for gradient fills, so the plot in Figure 13.1 is not possible in standard R graphics.

In this chapter, we will look at the **gridSVG** package, which provides access from R to advanced graphics features that R itself does not support.

```
> library(gridSVG)
```

13.1   Exporting **SVG**

The main function in the **gridSVG** package is the `grid.export()` function. This generates an SVG file from the current page of **grid** output. For example, the following code draws a **lattice** boxplot on the normal R graphics screen

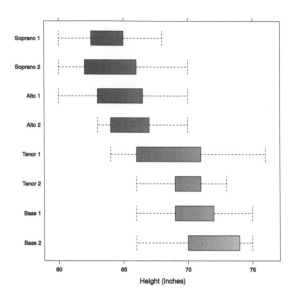

Figure 13.1
A **lattice** boxplot with the boxes filled using a linear gradient fill (from black to white).

device and then exports the plot to an SVG file with a call to `grid.export()` (see Figure 13.2).*

```
> library(lattice)
> bwplot(voice.part ~ height, data=singer,
          xlab="Height (inches)",
          par.settings=list(box.rectangle=list(col="black"),
                            box.umbrella=list(col="black"),
                            plot.symbol=list(col="black")))
> grid.export()
```

This functionality is not very useful in itself because the standard `svg()` graphics device can already generate SVG files. Furthermore, while the standard `svg()` device can export all R graphics output, **gridSVG** only exports **grid** output. The value of **gridSVG** lies in its functions that allow advanced SVG features to be added to a plot. Also, the **gridGraphics** package allows us to convert all base graphics into **grid** graphics (see Sections 12.2 and 13.7),

---

*The data used in this example are heights of singers, grouped by voice part, available as the data set **singers** in the **lattice** package.

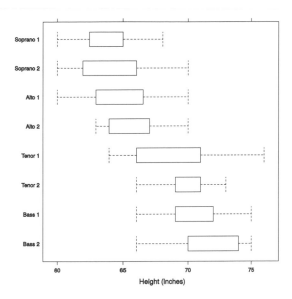

Figure 13.2
A **lattice** boxplot that has been exported to SVG using the `grid.export()` function.

which means that we can actually export base graphics with **gridSVG** with a tiny bit of extra work.

As a simple example of advanced SVG features, we will explore filling a rectangle with a linear gradient. The following **grid** code draws a simple rectangle with no fill color (see the left side of Figure 13.3).

```
> grid.rect(name="r")
```

The **gridSVG** package provides the function `linearGradient()` to describe a linear gradient fill and the function `grid.gradientFill()` to apply a gradient fill to a **grid** grob. The following code uses these functions to define a horizontal linear gradient from black to white then back to black and then apply that gradient fill to the rectangle that we just drew. Notice that we use the name of the rectangle, `"r"`, to identify the grob that we want to apply the gradient fill to.

```
> gradient <- linearGradient(c("black", "white", "black"),
                             x0=0, y0=.5, x1=1, y1=.5)
```

```
> grid.gradientFill("r", gradient)
```

Figure 13.3

On the left is a normal **grid** rectangle (with no fill); on the right is the same rectangle filled with a horizontal linear gradient fill (from black to white and back to black).

After running this code, absolutely nothing will have changed on the standard R graphics device; it will still look like the left side of Figure 13.3. This should not be surprising because normal R graphics does not support gradient fills. However, if we call the `grid.export()` function we will produce an SVG file that contains the gradient fill. The result is shown on the right side of Figure 13.3.

```
> grid.export()
```

By default, the result of a call to `grid.export()` is an SVG file with the name `"Rplots.svg"`, but we can specify a different file name as the first argument to `grid.export()` if we wish.

In addition to gradient fills, the **gridSVG** package provides access to pattern fills, clipping paths, opacity masks, and image filters, all of which are beyond the capabilities of standard R graphics devices. Table 13.1 provides a full list of the functions that can be used to define a special effect and the functions that can be used to apply the effect to a **grid** grob, and the following sections explore each of these features in more detail.

It is worth emphasizing the fact that the functions that apply an SVG special effect must identify a **grid** grob by name. This is a good thing in the sense that we can localize a special effect to only parts of an image and we can apply special effects *after* the image has been drawn. On the other hand, for this to work, we need the grobs in an image to be named. Fortunately, packages like **lattice** and **ggplot2** both do a reasonable job of naming most grobs in any image that they create. However, we have no guarantee that grobs created by someone else's code will be sensibly named, in which case applying SVG special effects to the correct part of an image can be challenging.

Table 13.1

SVG special effects that are available in the **gridSVG** package. In each case, there is a function to define the effect and a function to apply the effect to a **grid** grob.

Effect	Define	Apply
Linear gradient	`linearGradient()`	`grid.gradientFill()`
Radial gradient	`radialGradient()`	`grid.gradientFill()`
Clipping path	`clipPath()`	`grid.clipPath()`
Filter	`filterEffect()`*	`grid.filter()`
Opacity mask	`mask()`	`grid.mask()`
Pattern fill	`pattern()`	`grid.patternFill()`

*A filter effect is defined by one or more filters, such as a Gaussian blur filter defined by the `feGaussianBlur()` function.

13.2 **SVG** advanced features

This section demonstrates examples of each of the SVG special effects from Table 13.1.

13.2.1 Gradient fills

In addition to the linear gradients that were described in the previous section, we can also create and apply radial gradient fills. The `radialGradient()` function is used to describe the fill and `grid.gradientFill()` applies the fill to one or more grobs.

A radial gradient is defined by a set of colours for the gradient to transition through, plus a location for each colour in the set. The code below shows the simplest case, where we have a gradient from one colour to another (in this case white to black), with the first colour at the centre of the shape being filled and the second colour at the edge of the shape being filled. The shape we are filling is a rectangle (see Figure 13.4).

```
> grid.rect(name="r1")
> rg1 <- radialGradient(c("white", "black"))
> grid.gradientFill("r1", rg1)
```

The next example shows a slightly more complex radial gradient. This time we have three colours (white then black then white) and we have set the "focus" of the gradient to the left of center. We have also specified that the

 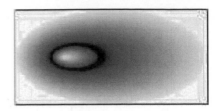

Figure 13.4
On the left is a rectangle with a simple radial gradient from white to black; on the right is a more complex radial gradient from white to black then back to white, with the focus of the gradient skewed to the left and the first transition happening faster than the second.

first transition from white to black occurs in the first quarter of the distance from the centre to the edge and the transition back to white occurs more slowly. The final result is a skewed gradient, again within a rectangle (see Figure 13.4).

```
> grid.rect(name="r2")
> rg2 <- radialGradient(c("white", "black", "white"),
                        stops=c(0, .25, 1),
                        fx=.25, fy=.5)
> grid.gradientFill("r2", rg2)
```

In both examples above, we apply the radial gradient to a specific grob by naming the grob in the first argument to the `grid.gradientFill()` function. It is also possible to treat the first argument as a regular expression, by specifying `grep=TRUE`, and to allow more than one grob to be affected by the fill, by specifying `global=TRUE`.

13.2.2   Pattern fills

Another way that we can fill shapes is with a pattern fill. These are created with the `pattern()` function and applied to one or more grobs with `grid.patternFill()`.

A pattern is based on a **grid** grob, for example, the following code defines a pattern based on a circle.

```
> dots <- pattern(circleGrob(r=.3, gp=gpar(fill="black")))
```

When a pattern fill is applied to another grob, the pattern is repeated to fill up the shape. We refer to the basic pattern as a *tile* and the pattern definition

Figure 13.5
On the left is a rectangle with a pattern fill based on a circle, where the pattern
starts at the bottom left of the rectangle and repeats 10 times in each direction. On
the right is a rectangle also filled with a pattern based on a circle, but this time the
pattern starts at the centre of the rectangle and repeats every 1cm.

includes how to repeat the tile to fill up a shape. By default, the tile starts off
centred at the bottom-left corner of the shape and repeats ten times across
and ten times up to fill the shape (see Figure 13.5).

```
> grid.rect(name="r1")
> grid.patternFill("r1", dots)
```

The following code defines a different pattern, still based on the simple circle,
but with the pattern starting in the centre of the filled shape and with a fixed
tile size of 1cm (see Figure 13.5).

```
> dotgrid <- pattern(circleGrob(r=.3, gp=gpar(fill="black")),
                     x=.5, y=.5,
                     width=unit(1, "cm"), height=unit(1, "cm"))
> grid.rect(name="r2")
> grid.patternFill("r2", dotgrid)
```

The definition of a pattern does not have to be based on just a single simple
grob. For example, the following code creates a pattern fill that is based on a
combination of gray rectangles and a white circle. This fill is then used on a
rectangle (see Figure 13.6).

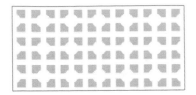 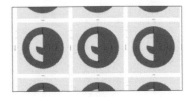

Figure 13.6

On the left is a pattern fill that is based on a combination of grobs (a white circle on top of four gray rectangles). On the right is a pattern fill that is based on a **ggplot2** plot.

```
> c <- circleGrob(r=.25, gp=gpar(col=NA, fill="white"))
> r <- rectGrob(x=c(1, 1, 3, 3)/4, y=c(1, 3, 3, 1)/4,
                width=.3, height=.3,
                gp=gpar(col=NA, fill="grey"))
> p <- pattern(gTree(children=gList(r, c)),
               x=.5, y=.5,
               width=unit(2, "cm"), height=unit(2, "cm"))
> grid.rect(name="r3")
> grid.patternFill("r3", p)
```

Just to show how general this approach is, the following code uses a **ggplot2** plot as the basis for a pattern fill (see Figure 13.6).

```
> library(ggplot2)
> cxc <- ggplot(mtcars, aes(x = factor(cyl))) +
         geom_bar(width = 1, colour = "black") +
         coord_polar(theta = "y")
> gg <- ggplotGrob(cxc)
> p <- pattern(gg, x=.5, y=.5,
               width=unit(4, "cm"), height=unit(4, "cm"))
> grid.rect(name="r4")
> grid.patternFill("r4", p)
```

13.2.3   Filters

A *filter* is a graphical operation that can be applied to a grob. A filter is defined by calling the `filterEffect()` function, which includes calls to specific filter effect functions such as `feGaussianBlur()`, then the filter is applied by calling `grid.filter()`.

Figure 13.7
On the left is a rectangle with a Gaussian blur filter applied to it (so that the border has become fuzzy). On the right is a rectangle with a complex filter applied. The filter takes the original rectangle, offsets it down and to the right, extracts the opaque section of the rectangle (which is all of the rectangle because it is filled with white), blurs the result, then composites the original rectangle over the top. The end result is a drop shadow.

The following code describes a simple filter effect, which applies a Gaussian blur to a rectangle (see Figure 13.7).

```
> feSimple <- filterEffect(feGaussianBlur(sd=3))
> grid.rect(name="r1", width=.8, height=.8)
> grid.filter("r1", feSimple)
```

The next code describes a more complex filter effect, which is a combination of several filters. The image being filtered is again just a simple rectangle. However, the first filter takes the alpha channel of the source image (and because the source image is opaque, that is the entire rectangle) and offsets it down and to the right. The result of this filter is given the label "offOut". The second filter takes the "offOut" filter result and blurs it; the result of this filter is given the label "gaussOut". The third filter composites the original image (the rectangle) with the "gaussOut" filter result to produce a final result that is a rectangle with a drop shadow (see Figure 13.7).

```
> offset <- feOffset("SourceAlpha", result="offOut",
                     dx=unit(2, "mm"), dy=unit(-2, "mm"))
> blur <- feGaussianBlur("offOut", sd=3, result="gaussOut")
> blend <- feBlend("SourceGraphic", "gaussOut")
> feComplex <- filterEffect(list(offset, blur, blend))
> grid.rect(name="r2", width=.8, height=.8,
            gp=gpar(fill="white"))
> grid.filter("r2", feComplex)
```

Table 13.2 lists the full set of filter effects that are supported by **gridSVG**.

Table 13.2

Filter effects supported by **gridSVG**

Filter	Description
feBlend	Blend two objects together.
feColorMatrix	Apply a matrix transformation on colour values.
feComponentTransfer	Perform colour component-wise remapping.
feComposite	Combine images using Porter-Duff operations.
feConvolveMatrix	Apply a matrix convolution filter effect.
feDiffuseLighting	Light an image using the alpha channel as a bump map.
feDisplacementMap	Displace pixel values from a filter input.
feDistantLight	Create a distant light source.
feFlood	Create and fill a rectangular region.
feGaussianBlur	Apply a Gaussian blur to an image.
feImage	Draw a referred image.
feMerge	Composite image layers together.
feMorphology	"Fatten" or "thin" artwork.
feOffset	Offset an input image relative to its current position.
fePointLight	Create a point light source.
feSpecularLighting	Light an image using the alpha channel as a bump map.
feSpotLight	Create a spot light source.
feTile	Fill a rectangle with a tiled pattern of an input image.
feTurbulence	Create an image using the Perlin turbulence function.

Figure 13.8
On the left is a rectangle with a linear gradient fill (like Figure 13.3) that has been clipped using a circle as the clipping path. On the right is the same rectangle that has been clipped using three overlapping circles as the clipping path.

13.2.4   Clipping paths

Standard R graphics allows for clipping of graphical output, but only to rectangular regions. With **gridSVG**, we are able to clip to arbitrary paths.

A clipping path is defined using the `clipPath()` function and the clipping is applied to a grob with the `grid.clipPath()` function. Similar to pattern fills, a clipping path is based upon a **grid** grob. For example, the following code draws a rectangle with a linear gradient fill (like in Figure 13.3), then applies a clipping path based on a circle (see Figure 13.8).

```
> grid.rect(name="r1")
> grid.gradientFill("r1", gradient)
> cp <- clipPath(circleGrob())
> grid.clipPath("r1", cp)
```

The following code shows that a clipping path can be made up from more than one shape. In this case, the clipping path is defined by three overlapping circles; the resulting clipping path is the union of the three circles (see Figure 13.8).

```
> grid.rect(name="r2")
> grid.gradientFill("r2", gradient)
> cp <- clipPath(circleGrob(x=1:3/4, r=.3))
> grid.clipPath("r2", cp)
```

13.2.5   Masks

A mask is similar to clipping in that it allows us to exclude parts of an image. However, where clipping describes an outline outside of which the image is

 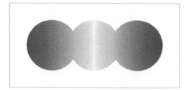

Figure 13.9
Applying a black-and-white mask. On the left is the mask itself (three overlapping white circles on a black background) and on the right is a rectangle with a linear gradient fill that has had the mask applied to it. Only the parts of the filled rectangle where the mask was white are visible.

discarded, masking allows degrees of exclusion.

A mask is itself an image that contains white, black, or gray components. When a mask is applied to another image, where the mask is white, the masked image is retained, where the mask is black, the masked image is discarded, and where the mask is gray, the masked image becomes semitransparent.

A mask is defined with the `mask()` function and applied to a grob with the `grid.mask()` function. The following code shows a simple case where the mask is a set of three white circles on a black background and the mask is applied to the rectangle with a linear gradient. The result is exactly the same as the second clipping example from the previous section (see Figure 13.9).

```
> circlesOnBlack <-
      gTree(children=gList(rectGrob(gp=gpar(fill="black")),
                           circleGrob(x=1:3/4, r=.3,
                                      gp=gpar(col=NA,
                                              fill="white"))))
> m <- mask(circlesOnBlack)


> grid.rect(name="r2")
> grid.gradientFill("r2", gradient)
> grid.mask("r2", m)
```

The following code creates a more sophisticated mask. In this case, we use the rectangle with a linear gradient as the mask and apply the mask to three white circles on a black background. Where the mask is gray, the masked image becomes semitransparent (see Figure 13.10).

Figure 13.10

Applying a grayscale mask. On the left is the mask itself (a rectangle filled with a grayscale linear gradient) and on the right is a set of three white circles on a black background that has had the mask applied to it. Where the mask is white, the circles and black background are opaque and where the mask is gray, the circles and black background are semitransparent.

```
> grayGradient <-
      gTree(children=gList(gradientFillGrob(rectGrob(),
                                            gradient)))
> m <- mask(grayGradient)

> masked <- maskGrob(circlesOnBlack, m)
> grid.draw(masked)
```

13.3 **SVG** drawing context

In addition to applying features to specific **grid** grobs, it is possible to add a clipping path or an opacity mask to the current drawing context so that it affects all subsequent drawing. This is the purpose of the `pushClipPath()` and `pushMask()` functions. These actions are similar to pushing a **grid** viewport in that they affect the drawing context, but only within the current viewport. The resulting clipping path or opacity mask is only in effect until the next call to `popViewport()`. We can also call `popClipPath()`, or `popMask()`, or `popContext()` to revert the drawing context without leaving the current viewport.

The following code is equivalent to the `grid.clipPath()` example that produced the left side of Figure 13.8. In the original example, we drew the rectangle, then filled it, then applied a clipping path to it. This time, we enforce a clipping path and then draw the rectangle and fill it (and then roll back the clipping path). The difference in the latter case is that we could draw more than just a single rectangle between the `pushClipPath()` call and the `popClipPath()` call.

```
> pushClipPath(path)
> grid.rect(name="r")
> grid.gradientFill("r", gradient)
> popClipPath()
```

13.4   **SVG** definitions

Clipping paths, opacity masks, filters, gradient fills, and pattern fills can be collectively referred to as SVG *definitions*. There are three steps involved in using one of these definitions: we create the definition, we register the definition, and we apply the definition. In the examples so far, we have only seen the first and the last step; the registering step has been handled automatically. For example, we have used `linearGradient()` to create a linear gradient fill definition and we have used `grid.gradientFill()` to apply that linear gradient.

We can explicitly register a definition before we use it with functions like `registerGradientFill()` or `registerClipPath()`. One reason for doing this is efficiency because registering only records the definition in the SVG file once, which will reduce the size of the SVG file; if we do not register the definition, a copy is added to the SVG file every time that we use it. A second reason for explicitly registering a definition is that the registration step is when the semantics of the definition are resolved. To demonstrate the concept of registration, we will consider the linear gradient that we described earlier. We can see that this gradient will smoothly transition from black, at location (0, 0.5), to white, at location (1, 0.5).

```
> gradient <- linearGradient(c("black", "white", "black"),
                             x0=0, y0=.5, x1=1, y1=.5)
```

What is the meaning of those locations? The default is that 0 is the left side of the object being filled and 1 is the right side, but, as the example at the start of this chapter shows (see Figure 13.1), we might also want to fill several objects with a single gradient that is defined relative to the page or a whole plot.

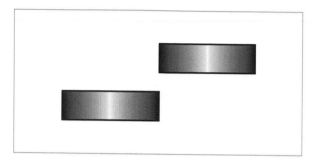

Figure 13.11
Two rectangles filled with a linear gradient that is defined relative to the bounding box of the object being filled; both rectangles receive the same fill.

To get the latter effect, we have to change two things. First of all, we have to specify the coordinate system that we are using to describe the gradient fill. We do this with the `gradientUnits` argument, which can have the value `"bbox"` (the default) or `"coords"`. We will demonstrate the difference through a series of examples.

The following code defines a linear gradient from black to white and back to black, that starts at $(0, .5)$ and ends at $(1, .5)$, where those locations are relative to the bounding box of the object being filled (the default).

```
> gradientBBox <- linearGradient(c("black", "white", "black"),
                                 gradientUnits="bbox",
                                 x0=0, y0=.5, x1=1, y1=.5)
```

In the next code, we apply that fill to two rectangles and, because the gradient is relative to each rectangle, both rectangles are filled the same way (see Figure 13.11).

```
> grid.rect(1:2/3, 1:2/3, width=1/3, height=.2, name="r2")
> grid.gradientFill("r2", gradientBBox)
> grid.export()
```

If we define the same gradient fill, but use `gradientUnits="coords"`, the gradient fill is defined relative to the whole page rather than the bounding box of the object being filled.

```
> gradientPage <- linearGradient(c("black", "white", "black"),
                                 gradientUnits="coords",
                                 x0=0, y0=.5, x1=1, y1=.5)
```

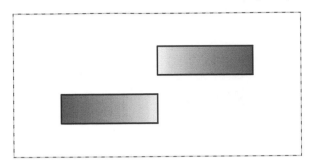

Figure 13.12
Two rectangles filled with a linear gradient that is defined relative to the whole page (as indicated by the dotted rectangle); each rectangle receives a different fill based on where it lies on the page.

If we fill two separate rectangles with this gradient fill, they receive different fills based on where the rectangle lies on the page (see Figure 13.12).

```
> grid.rect(1:2/3, 1:2/3, width=1/3, height=.2, name="r2")
> grid.gradientFill("r2", gradientPage)
> grid.export()
```

Finally, if we define a gradient fill using **gradientFill="coords"**, the locations in the definition are evaluated when the gradient fill is registered, which means that we can control where on the page the gradient fill is defined (e.g., within a plot region rather than over the entire page). For example, in the following code we will reuse the **pageGradient** from above. However, we first push a viewport that only occupies the central third of the page (a dotted rectangle is drawn to show where this viewport lies on the page; see Figure 13.13). We then explicitly register the gradient fill with a call to **registerGradientFill()**, which means that the gradient fill is defined across the central third of the page. Two separate rectangles are then filled with that gradient and the result is different for each rectangle (because they are in different positions on the page) and different from the previous example (because the gradient fill is not defined across the whole page). In the call to **grid.gradientFill()**, rather than providing a gradient fill object as in previous examples, we identify a registered gradient fill by its label (in this case, **"g"**). The final result is shown in Figure 13.13.

Outside of the central third of the page the gradient fill just stays black (the value at the edge of the gradient fill), but that behaviour can be controlled by the **spreadMethod** argument to the **linearGradient()** function.

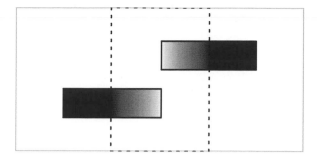

Figure 13.13
Two rectangles filled with a linear gradient that is defined relative to a viewport in
the central third of the page (as indicated by the dotted rectangle); each rectangle
receives a different fill based on where it lies relative to the viewport.

```
> pushViewport(viewport(width=1/3, name="vp"))
> registerGradientFill("g", gradientPage)
> upViewport()
> grid.rect(1:2/3, 1:2/3, width=1/3, height=.2, name="r2")
> grid.gradientFill("r2", label="g")
> grid.export()
```

13.5   Drawing off screen

The `grid.export()` function works by exporting whatever is on the current
graphics device to an SVG file. This requires that we first draw output on a
standard R graphics device, e.g., on screen, and then call `grid.export()`.

The `gridsvg()` function offers a different approach. We can use this like any
other R graphics device (e.g., `pdf()`), to start a **gridSVG** graphics device
and then nothing is drawn on the screen; we just end up with the SVG file
once we call `dev.off()` to close the **gridSVG** device.

It is also possible to avoid generating an external SVG file. If we set the
`name` argument to `grid.export()` to NULL then, instead of generating an SVG
file, the function returns a list, the first element of which is an in-memory
representation of the SVG code. This object is an `"XMLInternalNode"` from
the package **XML** and it can be used to fine-tune the raw SVG code (using
other functions from the **XML** package). For example, the following code
draws nothing on screen, but generates SVG code for an image and then

extracts the `<circle>` element from that SVG code.

```
> pdf(NULL)
> grid.circle()
> svg <- grid.export(NULL)$svg
> dev.off()

> library(XML)
> getNodeSet(svg, "//svg:circle",
             namespaces=c(svg="http://www.w3.org/2000/svg"))

[[1]]
<circle id="GRID.circle.469.1.1" cx="252" cy="252" r="252"/>

attr(,"class")
[1] "XMLNodeSet"
```

Most of the examples in this chapter have focused on drawing **grid** grobs
with specific names and then selecting which grob to modify by specifying
the grob name. As with standard **grid** functions, there are *Grob versions
of the **grid.*** functions, so we can also work directly with grobs off screen.
For example, the following code creates a rectangle grob, but does not draw
it, then adds a gradient fill effect to it, and finally draws the gradient-filled
rectangle.

```
> rect <- rectGrob()
> rectFilled <- gradientFillGrob(rect, gradient)
> grid.draw(rectFilled)
```

Section 13.2.5 also contains an example of this approach.

13.6   SVG fonts

In R graphics, we can specify a single font to use for a piece of text. For
example, in **grid** graphics, we do this by specifying a value for the `fontfamily`
argument in a call to **gpar()**, as shown below.

```
> grid.text("hello", gp=gpar(fontfamily="serif"))
```

In an SVG image, because SVG is used in web pages and those web pages can be viewed on many different machines, so we cannot know what fonts are available to the viewer, it is possible to specify a list of fonts for a piece of text. Whatever software is used to view the SVG file will go down the list of fonts until it finds one that it can use. The SVG code below shows an example, where the font Helvetica will be used if it is available, otherwise Arial will be used if it is available, otherwise a generic sans-serif font will be used.

```
<text style="font-family: Helvetica, Arial, sans-serif">
    hello
</text>
```

The **gridSVG** package calls this list of fonts a *font stack* and provides functions to define different font stacks. When we export an R plot that contains text, the font family specification for that text is used to select a font stack, which generates a list of fonts in the SVG file.

There are three font stacks available and the `getSVGFonts()` function returns the current settings.

```
> stacks <- getSVGFonts()
> stacks

$sans
[1] "Helvetica"       "Arial"          "FreeSans"
[4] "Liberation Sans" "Nimbus Sans L"  "sans-serif"

$serif
[1] "Times"
[2] "Times New Roman"
[3] "Liberation Serif"
[4] "Nimbus Roman No9 L Regular"
[5] "serif"

$mono
[1] "Courier"     "Courier New"   "Nimbus Mono L"
[4] "monospace"
```

The `setSVGFonts()` function can be used to modify the font stacks. The following code makes use of the font stacks and some of the ideas from previous sections to create an SVG image that makes use of a handwriting-style Google Font.*

---

*https://fonts.google.com/?category=Handwriting\&selection.family=Satisfy

The first step is to define a custom font stack. This font stack will use the
Google Font "Satisfy" if it is available, or otherwise any serif font that can be
found.

```
> stacks$serif <- c("Satisfy", "serif")
> setSVGFonts(stacks)
```

We now draw an image containing text and specify the "serif" font family,
which means that the serif font stack will be used. We do all drawing off
screen and just capture the SVG output for this image.

```
> pdf(NULL, width=2, height=1)
> grid.text("hello", gp=gpar(fontfamily="serif"))
> svg <- grid.export(NULL)$svg
> dev.off()
```

Next, we add a node into the SVG output to make sure that the web browser
that views the SVG will find the Google Font.

```
> root <-
      xmlRoot(svg, "svg:svg",
                  namespaces=c(svg="http://www.w3.org/2000/svg"))
> url <-
      "url('https://fonts.googleapis.com/css?family=Satisfy');"
> styleNode <-
      newXMLNode("style",
                    attrs=c(type="text/css"),
                    paste("@import", url))
> invisible(newXMLNode("defs", styleNode, parent=root))
```

The final step is to write the modified SVG code out to a file. The final image
is shown in Figure 13.14.

```
> saveXML(root, "Figures/export-fonts.svg")
```

```
[1] "Figures/export-fonts.svg"
```

One problem with the font handling in **gridSVG** output is that the fine
placement of text will not be accurate with non-standard fonts, because the
font metrics the R uses to position text are not necessarily the same as the
font metrics of the font that is ultimately viewed on screen.

Figure 13.14
An SVG image generated by **gridSVG** with text that uses a Google Font ("Satisfy").

13.7 Exporting base graphics

Although, as the name suggests, **gridSVG** is limited to exporting **grid** output, the **gridGraphics** package allows us to translate any base graphics output to **grid**, so we can in effect export any R graphics.

The following code sequence shows the steps involved. First, we draw a base plot ...

```
> plot(mpg ~ disp, mtcars)
```

... then we convert the base plot to **grid** graphics ...

```
> library(gridGraphics)
> grid.echo()
```

... then, perhaps after adding special SVG features to the plot grobs, we export the plot to an SVG file ...

```
> library(gridSVG)
> grid.export()
```

13.8 Exporting to other formats

A major limitation of the **gridSVG** package is that it can only produce SVG output. This is an excellent graphics format for including images in web pages, or HTML documents generally, but it is not appropriate for including images within, for example, PDF reports produced from LaTeX documents.

If we wish to produce an R plot that contains SVG special effects, but in another format, like PDF, we need to convert from the SVG that `grid.export()` produces to the format that we desire. Fortunately, there are several programs that can perform this conversion, though the quality of the result can vary between programs. For example, some programs will produce a raster result when converting SVG files with special effects like gradient fills to PDF files.

Inkscape is one program that performs quite well at this task, with the added advantage that it is available on Windows as well as Linux. On Linux it is also relatively straightforward to write code for this task, which allows the conversion to be included in an automated workflow. The print version of this book has exactly this problem and solves it with code along the lines of the code below.

```
> system("inkscape --export-pdf=output-file.pdf input-file.svg")
```

Another option is to use a "headless" browser, as demonstrated by the following code.

```
> system("chromium-browser --headless
          --print-to-pdf input-file.svg")
```

13.9   Exporting imported images

Section 11.3.2 described the **grImport2** package for importing SVG images into R. One feature of that package is that it can import SVG images that contain features that standard R graphics cannot support.

The **gridSVG** package can be used in combination with **grImport2** to generate SVG output that contains imported images that R by itself cannot render. An example of this idea is shown in Figure 11.14.

Chapter summary

The **gridSVG** package exports **grid** output to an SVG format. The advantage of this over the standard SVG graphics device is that advanced SVG features can be added to an image and exported. These features include gradient fills, pattern fills, filters, clipping paths, and masks. Base graphics output can be exported by first converting to **grid** output with the **gridGraphics** package.

Bibliography

Daniel Adler and Duncan Murdoch. *rgl: 3D Visualization Device System (OpenGL)*, 2010. R package version 0.91.

Adobe Systems Inc. *PostScript Language Reference Manual*. Addison-Wesley Longman, 2nd edition, 1990.

JJ Allaire, Yihui Xie, Jonathan McPherson, Javier Luraschi, Kevin Ushey, Aron Atkins, Hadley Wickham, Joe Cheng, and Winston Chang. *rmarkdown: Dynamic Documents for R*, 2017. URL `https://CRAN.R-project.org/package=rmarkdown`. R package version 1.8.

Richard A. Becker and John M. Chambers. *Extending the S System*. Chapman & Hall, 1985.

Richard A. Becker, William S. Cleveland, and Ming-Jen Shyu. The visual design and control of trellis display. *Journal of Computational and Graphical Statistics*, 5:123–155, 1996.

Richard A. Becker, Allan R. Wilks, Ray Brownrigg, Thomas P Minka, and Alex Deckmyn. *maps: Draw Geographical Maps*, 2018. URL `https://CRAN.R-project.org/package=maps`. R package version 3.3.0.

Ray Brownrigg, Richard A. Becker, and Allan R. Wilks. *mapdata: Extra Map Databases*, 2018. URL `https://CRAN.R-project.org/package=mapdata`. R package version 2.3.0.

J. M. Chambers. Structured computational graphics for data analysis. *Proceedings of the International Statistical Institute*, 40:501–507, 1975.

Winston Chang. *extrafont: Tools for using fonts*, 2014. URL `https://CRAN.R-project.org/package=extrafont`. R package version 0.17.

Winston Chang, Alexej Kryukov, and Paul Murrell. *fontcm: Computer Modern font for use with extrafont package*, 2014. URL `https://CRAN.R-project.org/package=fontcm`. R package version 1.1.

William S. Cleveland. *The Elements of Graphing Data*. Wadsworth Publ. Co., 1985.

William S. Cleveland. *Visualizing Data*. Hobart Press, 1993.

William S. Cleveland and Robert McGill. Graphical perception: The visual decoding of quantitative information on graphical displays of data. *Journal of the Royal Statistical Society, Series A, General*, 150:192–210, 1987.

P. Dalgaard. *Introductory Statistics with R*. Statistics and Computing. Springer New York, 2008. ISBN 9780387790534. URL `https://books.google.co.nz/books?id=YIOkT8cuiVUC`.

Matthew W. Felgate, Simon H. Bickler, and Paul R. Murrell. Estimating parent population of pottery vessels from a sample of fragments: a case study from inter-tidal surface collections, roviana lagoon, solomon islands. *Journal of Archaeological Science*, 40(2):1319 – 1328, 2013.

John Fox. *An R and S-Plus Companion to Applied Regression*. Sage Publications, 2002.

Michael Friendly. *Visualizing Categorical Data*. SAS Publishing, 2000.

David Gohel, Hadley Wickham, Lionel Henry, and Jeroen Ooms. *gdtools: Utilities for Graphical Rendering*, 2018. URL `https://CRAN.R-project.org/package=gdtools`. R package version 0.1.7.

J.A. Hartigan and B. Kleiner. A mosaic of television ratings. *The American Statistician*, 38:32–35, 1984.

Richard M. Heiberger and Burt Holland. *Statistical Analysis and Data Display: An Intermediate Course with Examples in S-PLUS, R, and SAS*. Springer, 2004.

A.V. Hershey. A contribution to computer typesetting techniques: Tables of coordinates for Hershey's repertory of occidental type fonts and graphic symbols. *NBS Special Publication 424*, April 1976.

H. Hofmann and M. Theus. Interactive graphics for visualizing conditional distributions. Unpublished manuscript, 2005.

Torsten Hothorn, Kurt Hornik, and Achim Zeileis. Unbiased recursive partitioning: A conditional inference framework. *Journal of Computational and Graphical Statistics*, 15(3):651–674, 2006.

J. Hummel. Linked bar charts: Analyzing categorical data graphically. *Computational Statistics*, 11:23–33, 1996.

Ross Ihaka, Paul Murrell, Kurt Hornik, and Achim Zeileis. *colorspace: Color Space Manipulation*, 2016. R package version 1.3-2.

L. Kaufman and P.J. Rousseeuw. *Finding Groups in Data: An Introduction to Cluster Analysis*. Wiley, New York, 1990.

O.P. Lamigueiro. *Displaying Time Series, Spatial, and Space-Time Data with R*. Chapman & Hall/CRC The R Series. CRC Press, 2014. ISBN 9781466565227.

Duncan Temple Lang and the CRAN Team. *XML: Tools for Parsing and Generating XML Within R and S-Plus*, 2018. URL `https://CRAN.R-project.org/package=XML`. R package version 3.98-1.11.

H. W. Lie and B. Bos. *Cascading Style Sheets, Level 1*, 1996. W3C Recommendation.

Martin Maechler, Peter Rousseeuw, Anja Struyf, Mia Hubert, and Kurt Hornik. *cluster: Cluster Analysis Basics and Extensions*, 2018. R package version 2.0.7-1.

John Maindonald and John Braun. *Data Analysis and Graphics Using R: An Example-Based Approach*. Cambridge University Press, 2003.

Doug McIlroy, Ray Brownrigg, Thomas P Minka, and Roger Bivand. *mapproj: Map Projections*, 2018. URL `https://CRAN.R-project.org/package=mapproj`. R package version 1.2.6.

A.E. Miller. The analysis of unreplicated factorial experiments from a geometric perspective. *Canadian Journal of Statistics*, 31:311–327, 2003.

Paul Murrell. Integrating grid graphics output with base graphics output. *R News*, 3(2):7–12, 2003.

Paul Murrell. Importing vector graphics: The grImport package for R. *Journal of Statistical Software*, 30(4):1–37, 2009.

Paul Murrell. *gridBase: Integration of base and grid graphics*, 2014. URL `https://CRAN.R-project.org/package=gridBase`. R package version 0.4-7.

Paul Murrell. The gridGraphics Package. *The R Journal*, 7(1): 151–162, 2015. URL `https://journal.r-project.org/archive/2015/RJ-2015-012/index.html`.

Paul Murrell and Velvet Ly. *gridDebug: Debugging 'grid' Graphics*, 2015. URL `https://CRAN.R-project.org/package=gridDebug`. R package version 0.5-0.

Paul Murrell and Simon Potter. *gridSVG: Export 'grid' Graphics as SVG*, 2017. URL `https://CRAN.R-project.org/package=gridSVG`. R package version 1.6-0.

Paul Murrell and Zhijian Wen. *gridGraphics: Redraw Base Graphics Using 'grid' Graphics*, 2018. URL `https://CRAN.R-project.org/package=gridGraphics`. R package version 0.3-0.

Kurt Nassau, editor. *Color for Science, Art and Technology*. Elsevier, 1998.

Erich Neuwirth. *RColorBrewer: ColorBrewer palettes*, 2014. R package version 1.1-2.

Jeroen Ooms. *magick: Advanced Graphics and Image-Processing in R*, 2018a. URL https://github.com/ropensci/magick. R package version 1.8.

Jeroen Ooms. *rsvg: Render SVG Images into PDF, PNG, PostScript, or Bitmap Arrays*, 2018b. URL https://CRAN.R-project.org/package=rsvg. R package version 1.3.

Andrea Peters and Torsten Hothorn. *ipred: Improved Predictors*, 2017. URL https://CRAN.R-project.org/package=ipred. R package version 0.9-6.

Simon Potter. *grImport2: Importing 'SVG' Graphics*, 2018. R package version 0.1-4.

Yixuan Qiu. *showtext: Using Fonts More Easily in R Graphs*, 2018. URL https://CRAN.R-project.org/package=showtext. R package version 0.5-1.

Thomas Rahlf. *Data Visualisation with R.* Springer International Publishing, New York, 2017. ISBN 978-3-319-49750-1. URL http://www.datavisualisation-r.com.

Naomi Robbins. *Creating More Effective Graphs*. Wiley, 2005.

P.J. Rousseeuw. A visual display for hierarchical classification. In E. Diday, Y. Escoufier, L. Lebart, J. Pages, Y. Schektman, and R. Tomassone, editors, *Data Analysis and Informatics 4*, pages 743–748. North-Holland, Amsterdam, 1986.

Jeffrey A. Ryan and Joshua M. Ulrich. *quantmod: Quantitative Financial Modelling Framework*, 2018. URL https://CRAN.R-project.org/package=quantmod. R package version 0.4-13.

Deepayan Sarkar. *Lattice: Multivariate Data Visualization with R.* Springer, New York, 2008.

Charlie Sharpsteen and Cameron Bracken. *tikzDevice: R Graphics Output in LATEX Format*, 2018. URL https://CRAN.R-project.org/package=tikzDevice. R package version 0.11.

A. Struyf, M. Hubert, and P.J. Rousseeuw. Integrating robust clustering techniques in S-PLUS. *Computational Statistics and Data Analysis*, 26: 17–37, 1997.

E. R. Tufte. *The Visual Display of Quantitative Information.* Graphics Press, 1989.

Edward R. Tufte. *Envisioning Information.* Graphics Press, 1990.

A. Unwin. *Graphical Data Analysis with R.* Chapman & Hall/CRC The R Series. CRC Press, 2015. ISBN 9781498715249.

Simon Urbanek. *png: Read and write PNG images*, 2013a. URL https:

//CRAN.R-project.org/package=png. R package version 0.1-7.

Simon Urbanek. *tiff: Read and write TIFF images*, 2013b. URL https://CRAN.R-project.org/package=tiff. R package version 0.1-5.

Simon Urbanek. *jpeg: Read and write JPEG images*, 2014. URL https://CRAN.R-project.org/package=jpeg. R package version 0.1-8.

J. Verzani. *Using R for Introductory Statistics, Second Edition*. Chapman & Hall/CRC The R Series. Taylor & Francis, 2014. ISBN 9781466590731. URL https://books.google.co.nz/books?id=O86uAwAAQBAJ.

Hadley Wickham. *ggplot2: Elegant Graphics for Data Analysis*. Springer, 2nd edition, 2016.

Leland Wilkinson. *The Grammar of Graphics*. Springer, 2nd edition, 2005.

Kevin Wright. *pals: Color Palettes, Colormaps, and Tools to Evaluate Them*, 2018. URL https://CRAN.R-project.org/package=pals. R package version 1.5.

Yihui Xie. *Dynamic Documents with R and knitr*. Chapman and Hall/CRC, Boca Raton, Florida, 2nd edition, 2015. URL https://yihui.name/knitr/. ISBN 978-1498716963.

Achim Zeileis and Gabor Grothendieck. zoo: S3 infrastructure for regular and irregular time series. *Journal of Statistical Software*, 14(6):1–27, 2005. doi: 10.18637/jss.v014.i06. URL http://www.jstatsoft.org/v14/i06/.

Index

Printed and bound by CPI Group (UK) Ltd, Croydon, CR0 4YY

24/10/2024

01778279-0016